The Mind of the Mathematician

THE JOHNS HOPKINS UNIVERSITY PRESS } BALTIMORE

Michael Fitzgerald and Ioan James

THE MIND OF THE MATHEMATICIAN

9 8 7 6 5 4 3 2 1

The Johns Hopkins University Press
2715 North Charles Street
Baltimore, Maryland 21218-4363

www.press.jhu.edu

Library of Congress Cataloging-in-Publication Data
Fitzgerald, Michael, 1946–
The mind of the mathematician / Michael Fitzgerald and Ioan
James.
    p. cm.
Includes bibliographical references and index.
ISBN-13: 978-0-8018-8587-7 (hardcover : acid-free paper)
ISBN-10: 0-8018-8587-6 (hardcover : acid-free paper)
1. Mathematicians—Psychology. 2. Mathematics—Psychological
aspects. 3. Mathematical ability. 4. Mathematical ability—Sex
differences. I. James, I. M. (Ioan Mackenzie), 1928– II. Title.
BF456.N7F58 2007
510.1′9—dc22   2006025988

A catalog record for this book is available from the British Library.

Contents

Preface

Psychologists have long been fascinated by mathematicians and their world. In this book we start with a tour of the extensive literature on the psychology of mathematicians and related matters, such as the source of mathematical creativity. By limiting both mathematical and psychological technicalities, or explaining them when necessary, we seek to make our review of research in this field easily readable by both mathematicians and psychologists. In the belief that they might also wish to learn about some of the human beings who helped to create modern mathematics, we go on to profile twenty well-known mathematicians of the past whose personalities we find particularly interesting. These profiles serve to illustrate our tour of the literature.

Among the many people we have consulted in the course of writing this book we would particularly like to thank Ann Dowker, Jean Mawhin, Allan Muir, Daniel Nettle, Brendan O'Brien, Susan Lantz, and Mikhail Treisman.

We also wish to thank Ohio University Press, Athens, Ohio (www.ohio .edu/oupress), for granting permission to reprint an excerpt from Don H. Kennedy's biography of Sonya Kovalevskaya, *Little Sparrow: A Portrait of Sonya Kovalevskaya*.

Introduction

Mathematics, according to the Marquis de Condorcet, is the science that yields the most opportunity to observe the workings of the mind. Its study, he wrote, is the best training for our abilities, as it develops both the power and the precision of our thinking. Henri Poincaré, in his famous 1908 lecture to the Société de Psychologie in Paris, observed that mathematics is the activity in which the human mind seems to take least from the outside world, in which it seems to act only of itself and on itself. He went on to describe the feeling of the mathematical beauty of the harmony of numbers and forms, of geometric elegance—the true aesthetic feeling that all real mathematicians know. According to the British mathematical philosopher Bertrand Russell (1910), mathematics possesses "not only truth but supreme beauty—a beauty cold and austere . . . yet sublimely pure, and capable of a stern perfection such as only the greatest art can show." In the words of Courant and Robbins (1941): "Mathematics as an expression of the human mind reflects the active will, the contemplative reason, and the desire for aesthetic perfection. Its basic elements are logic and intuition, analysis and construction, generality and individuality. Though different traditions may emphasize different aspects, it is only the interplay of these aesthetic forces and the struggle for their syntheses that constitute the life, usefulness, and supreme value of mathematical science." The modern French mathematician Alain Connes, in Changeux and Connes (1995), tells us that "exploring the geography of mathematics, little by little the mathematician perceives the contours and structures of an incredibly rich world. Gradually he develops a sensitivity to the notion of simplicity that opens up access to new, wholly unsuspected regions of the mathematical landscape."

Nearly seventy years, ago the Scottish-American mathematician Eric Temple Bell set out to write about mathematicians in a way that would grip the imagination. In the introduction to his immensely readable *Men of Mathematics*, first published in 1937, he begins by emphasizing, "The lives of mathematicians presented here are addressed to the general reader and to others who might wish to see what sort of human beings the men were who created modern mathematics." Unfortunately, it leaves the impression that many of the more notable mathematicians of the past were self-serving and quarrelsome. This is partly because Bell selected his subjects accordingly, but also because he was not above distorting the facts to make a good story. He was a man of strong opinions, not simply reflecting the prejudices of a bygone age.

While the history of mathematics goes back thousands of years, psychology, in the modern sense, only originated in the nineteenth century. A special interest in mathematicians was present from the early years of the subject. The Leipzig neurologist Paul Julius Möbius (grandson of the mathematician August Ferdinand Möbius), sought evidence of diagnostic categories that might be related to the creative behavior of mathematicians in his *Uber die Anlage zur Mathematik* ("on the gift for mathematics") of 1900. A little later the Swiss psychologists Edouarde Claparède and Théodore Flournoy organized an inquiry into mathematicians' working methods, while in 1906 Poincaré gave the seminal lecture on mathematical invention mentioned earlier. Psychoanalysts have displayed relatively little interest in mathematicians, although they have a lot to say about creativity in general, with illustrations mainly from the arts. As Storr points out in his well-known book *The Dynamics of Creation* (1972), psychologists are primarily concerned with the causes that lie behind creativity, rather than the reasons that drive those who enjoy creative gifts to make full and effective use of them. The word *genius* often occurs in the literature, usually meaning much the same as "exceptionally able." *Some Reflections on Genius*, by Lord Russell Brain (1960), provides a good introduction; other works are listed in the bibliography. The term has gradually changed in meaning over a long period, stretching back to classical times. Today the meaning seems rather uncertain, and we avoid using it ourselves.

The material in the tour of the literature that forms the first part of this book is of necessity somewhat miscellaneous in character, but we have arranged the different topics under three broad chapter headings. In the first chapter we describe the special attraction of mathematics and its distinctive culture. To counter the popular impression that mathematicians are not interested in anything else, we give some examples of mathematicians who contributed to music and the other arts. For the second part of this book, we are dependent on reliable biographical information. Fortunately some excellent biographies of mathematicians have been written, and we mention some of these.

Another popular misconception is that mathematicians spend their time making calculations. While this is not the case in general, there are some exceptional individuals, including a few mathematicians, who have an extraordinary ability to do so. These "lightning calculators," who are known as savants, have intrigued psychologists for many years. Alfred Binet's classic account of his investigations of two lightning calculators and others with savant skills, *Psychologie des grands calculateurs et joueurs d'échecs* ("psychology of the great calculators and chess-players") of 1894, is still worth reading. We summarize what is known about them and then go on to discuss various other phenomena that have led some investigators to believe that there may be modules in the brain that specialize in calculation.

In chapter two we present a summary of the vast literature on mathematical education, and we also discuss the vexed question of gender difference. Child prodigies sometimes emerge in mathematics, as they do in music and languages, although they do not always develop into full-fledged mathematicians. We give some striking examples. Finally we review the literature on the decline of productivity with age. This is well established across a wide range of activities; we present evidence to support the general belief that mathematics is no exception.

In the last chapter of the first part of this book, we turn to the fascinating problem of mathematical creativity and the role of intuition: intuition is perception via the unconscious, according to Carl Jung. There is no better place to start than the well-known memoir, *The Psychology of Invention in the Mathematical Field* (1945), by Poincaré's disciple Jacques Hadamard, which takes up many of the interesting questions raised by Poincaré in his famous lecture. At the same time that this book appeared, the Viennese psychiatrist Hans Asperger published a thesis in which he gave a clinical description of the mild form of autism to which his name has been attached, and this has turned out to have an important bearing on the study of creativity, especially in mathematics.

Asperger syndrome, as the disorder is called, is much more common than classical autism. In fact, as the psychologist Rosemary Dinnage (2004) has observed, "perhaps the current interest in Asperger syndrome springs from some withdrawn but unrecognized streak in all of us." People with the syndrome are attracted to mathematics and kindred subjects; indeed, such people are so common in the mathematical world that they pass almost unnoticed. Asperger believed that for highly intelligent people a trace of autism could be essential to success in the arts and sciences. He believed that the perseverance, drive for perfection, good concrete intelligence, ability to disregard social conventions, and unconcern about the opinions of others could all be seen as advantageous and possibly prerequisite for certain kinds of new thinking and creativity. In his book *Autism and Creativity*, Fitzgerald (2004) reviews the evidence for a link between certain mild forms of autism, especially Asperger syndrome, and creativity in various fields.

In his thesis of 1944, Asperger himself wrote about "autistic intelligence" and saw it as a sort of intelligence hardly touched by tradition and culture— "unconventional, unorthodox, strangely 'pure' and original, akin to the intelligence of pure creativity." In recent years, tests on people with autism have revealed autistic intelligence to be linguistic, spatial, musical, and logical in character; such individuals tend to be fascinated by abstraction and logic. Many features of Asperger syndrome enhance creativity, but the ability to focus intensely on a topic and to take endless pains is characteristic. People with Asperger syndrome live very much in their intellects, and certain forms of creativity benefit greatly from this, particularly mathematical creativity.

It is well established that mild forms of mood disorders, such as bipolar disorder, can be conducive to creativity in many fields, including mathematics. In the case of literature the connection with bipolar disorder is particularly strong, as Jamison (1993a) has shown in *Touched with Fire*. Hershman and Lieb (1998), in *Manic Depression and Creativity*, argue that this extends to science. We discuss the situation for mathematics, and again give examples in the second part of our book.

Part 2 of this book consists of profiles of twenty famous mathematicians of the past, arranged chronologically by date of birth. We have searched the literature for people whose personalities show features that illustrate the points we raise in Part 1. Some of them were quite normal (what psychologists call "neurotypical"). Some of them suffered from mood disorders, such as bipolar disorder, but most suffered from a developmental disorder, in particular, Asperger syndrome. It is always difficult to make retrospective diagnoses, but in some cases at least it seems likely that these disorders were present and played an important role in the lives of the individuals in question, with interesting implications for our understanding of mathematical creativity. There is a problem here in that some authorities, such as the well-known neurologist Oliver Sacks, regard attempts to make diagnoses on the basis of purely historical evidence with great suspicion. Nevertheless there is a tradition, to which we adhere, of venturing such diagnoses in appropriate cases, while emphasizing that for obvious reasons they cannot be as secure as those clinicians make in the cases of living persons. Further discussion of some of the issues involved can be found in Bunyan's *Life Histories and Psychobiography* (1984).

Part I

TOUR OF THE LITERATURE

Chapter 1 } Mathematicians and Their World

The Attraction of Mathematics

A mathematician, according to the Oxford English Dictionary, is someone who is skilled in mathematics. Such a person may be involved in teaching or research, or in applications of the discipline. He or she may use mathematics to make a living, or enjoy it more as a form of recreation. Until relatively recently, mathematics embraced much of what we now call natural science, and most mathematicians were actively interested in its applications even if they mainly worked on the pure side of the discipline. Some, however, regarded mathematics as an end in itself, and there is a tendency in the literature to write as if that were the general belief. In this book, we use the term "mathematics" in the broad sense, to include both pure and applied mathematics, and to some extent we also include mathematical physics, statistics, and computer science.

It is difficult to estimate the number of mathematicians in the world today, but there must be well over a million. Only a minority, perhaps fifty thousand, could be described as active in research, and of these only a much smaller minority, perhaps five thousand, are publishing research of lasting value. Some of them—not very many—are truly creative mathematicians, who become famous primarily because of their excellence in research. It is these creative mathematicians who loom so large in the history of mathematics, as they will in this book, but we must not forget that they form only a small, atypical minority of the many who describe themselves as mathematicians. We must be careful not to make sweeping statements about all mathematicians on the basis of what we know about this minority. We must also be careful not to make generalizations on the basis of what we know about pure mathematicians, especially that select group of specialists in the theory of numbers. Currently some eighty or ninety thousand research papers in the mathematical sciences are published every year, but the majority of these are not in pure mathematics.

Ordinary mathematicians do not care much about philosophical questions; they leave that to the mathematical philosophers. However, if challenged, they might at least be willing to say whether they are Platonists or not. Platonists believe that in addition to objects, there exists a world of concepts to which we have access by intuition. For mathematical Platonists, numbers exist independently of ourselves in some objective sense. The Platonist mathematician tries to *discover* properties that numbers already have. The non-Platonist regards numbers as entirely a construction of the human mind, endowed with properties that can be investigated, so that the aim of research is to create or invent mathematics rather than discover it. The debate between the two points of view goes back to classical times. The philosophical literature defending and attacking Platonism in mathematics is too vast for us to pursue the matter further here, but the reader may wish to consult the discussion about the types of reality of mathematical entities in Changeux and Connes (1995).

As mathematics continues to develop, it becomes increasingly difficult to capture its nature in a single definition, but people keep trying. The British mathematician G. H. Hardy described mathematicians as makers of patterns of ideas. A similar point of view was expressed by the mathematical philosopher A. N. Whitehead (1911) when he wrote: "The notion of the importance of patterns is as old as civilisation . . . mathematics is the most powerful technique for the understanding of patterns, and for the analysis of the relation of patterns." Mathematics, he maintained, defines and gives names to these patterns, which generally originate in the physical world, so that we can manipulate them in our own minds and communicate ideas about them more easily. Other ideas about the nature of mathematics are discussed in the literature, but none of them seem entirely satisfactory. Some of them fail to take into account the role of the mathematician in the establishment of mathematical knowledge. Nor do they allow for the development of our knowledge of mathematics over time. Again, the reader may wish to consult the book by Changeux and Connes (1995).

Mathematics, like other sciences, advances by correcting and re-correcting mistakes, but as the French mathematician Elie Cartan (1952–1955) explained, "A desire to avoid mistakes forces mathematicians to find and isolate the essence of the problems and entities considered. Carried to an extreme, this procedure justifies the well-known joke according to which a 'mathematician is a scientist who neither knows what he is talking about nor whether whatever he is indeed talking about exists or not.'" The study of the foundations of mathematics is full of theories that turned out to be inconsistent and confident assertions that turned out to be wrong. We have come a long way from the a priori, human-independent conception of mathematics developed in nineteenth-century Germany and further developed in the twentieth century. Mathematics is no longer seen as transcendental, abstract, disembodied, unique, and independent of anything physical. On the contrary, it forms a part of human culture, a product of the human body, brain, and mind and of our experience in the physical world.

According to mathematical historian Morris Kline (1980):

mathematics creates by insight and intuition. Logic then sanctions the conquests of intuition. It serves as the hygiene that mathematics practises to keep its ideas healthy and strong. Moreover, the whole structure rests fundamentally on uncertain ground, the intuition of humans. Here and there an intuition is scooped out and replaced by a firmly built pillar of thought; however, this pillar is based on some deeper, perhaps less clearly defined, intuition. Though the process of replacing intuitions with precise thoughts does not change the nature of the ground on which mathematics ultimately rests, it does add strength and height to the structure.

By the second half of the nineteenth century, a great deal of thought was being given to the foundations of mathematics, especially in Germany. It was believed that it should be possible to find a system of axioms for mathematics that is demonstrably consistent and complete, in the sense that any definite statement within the system could always be either proved or disproved *within the same system*. The Viennese logician Kurt Gödel showed in 1930 that this was impossible. The modern view is to question whether mathematics needs foundations at all. Why not focus on what mathematicians actually do rather than attempt a description of the final product? There is an enormous amount of activity in the discipline, and the provision of rigorous proofs for theorems that are already well understood comes low on the agenda. In writing up work for publication, which may take months, the logical steps are provided, but they give a false impression of the way the result in question was obtained. Research involves imagination, experimentation, and a variety of other skills. Even incorrect arguments frequently lead to important insights. Mathematics is not a deductive science. "When you try to prove a theorem," says Paul Halmos (1985), "you don't just list the hypotheses and start to reason. What you do is trial and error, experimentation, guesswork." This is not a new idea; the British scientist Augustus de Morgan, in the mid-nineteenth century, declared that the moving power of mathematical invention is not reasoning but imagination.

What we say about mathematicians is almost equally true of many natural scientists, but every science has its own culture, and that of mathematics is distinctive. It is often difficult to decide whether someone is better regarded as a physicist or a mathematician. The cultural difference between the two disciplines has perhaps been best explained by the German topologist Max Dehn. In an address he gave to the faculty and students of the University of Frankfurt in 1928, entitled "The Mentality of the Mathematician," he said:

> I wish to say that, contrary to a widespread notion, the mathematician is not an eccentric, at any rate he is not eccentric because of his science. He stands between areas of study, especially between the humanities and the natural sciences, spheres that are unfortunately disjoint in our country. His method is only a version of the general scientific method. In virtue of the importance of the principle of the excluded middle, it is related to the juridical method. The object of his research is more spiritual than that of the natural scientist, and more sentient than that of the humanist. Connection with the natural sciences goes beyond the applications that permeate all the exact natural sciences. The mathematician knows that he owes to the natural sciences his most important stimulations. At times the mathematician has the passion of a poet or a conqueror, the rigour of his arguments is that of a responsible statesman or more simply, of a concerned father, and his tolerance and resignation are those of an old sage; he is revolutionary and conservative, sceptical and yet faithfully optimistic.

As David Hilbert explained in a radio broadcast on 8 September 1930:

> The tool which serves as intermediary between theory and practice, between thought and observation, is mathematics; it is mathematics which builds the linking bridges and gives the ever more reliable forms. From this it has come

about that our entire contemporary culture, inasmuch as it is based on the intellectual penetration and the exploitation of nature, has its foundations in mathematics. Already Galileo said: one can understand nature only when one has learned the language and the signs in which it speaks to us; but this language is mathematics and these signs are mathematical figures.

Although Albert Einstein is primarily regarded a physicist, some of his observations are often highly relevant to our theme, such as the following:

How can it be that mathematics, being after all a product of human thought which is independent of experience, is so admirably appropriate to the objects of reality? Is human reason, then, without experience, merely by taking thought, able to fathom the properties of real things? In my opinion the answer to this question is briefly this: as far as the propositions of mathematics refer to reality, they are not certain; as far as they are certain, they do not refer to reality. (Quoted in Schilpp 1951)

Several books have been written based on interviews with prominent mathematicians of today in which questions were asked such as, "What attracts you to mathematics?" Similar questions have been asked in surveys. Various answers have been given. In an interview, William Dunham said that to him, "The material world begins to seem so trivial, so arbitrary, so ephemeral when contrasted with the timeless beauty of mathematics" (Albers et al. 1997). When asked what pure mathematics was good for, Serge Lang replied, "It's good to give chills in the spine to a certain number of people, me included. I don't know what else it is good for, and I don't care" (see Lang 1985). This is the answer a pure mathematician might give, but other kinds of mathematicians might not agree with such an extreme view. A mathematical physicist, for example, might prefer to quote Eugene Wigner's (1960) aphorism about the extraordinary effectiveness of the application of mathematics to the real world.

Is there a tendency for mathematicians to set themselves apart? Mathematicians are not the loners they are often portrayed to be. Today about half of the research papers published in the mathematical sciences have more than a single author, although collaborative work was much less common in the past. A striking example is the list of over five hundred mathematicians who collaborated with the Hungarian mathematician Paul Erdös. If we extend the list to include mathematicians who collaborated with someone who collaborated with Erdös, we obtain almost seven thousand names. Usually mathe-

maticians who are interested in a particular research topic, such as sin-
gularities, establish a kind of invisible college in a completely informal way.
Those who belong to it may be the former research students, or "grandstu-
dents" (students of former students) and so on, of one of the major authorities on the subject, probably the person who invented it. They may be
scattered around the world, meet occasionally at international conferences,
and keep in touch by e-mail or through a newsletter. So even if the mathematician is the only person in his or her institution who is especially interested in a particular topic, he or she has no need to feel alone professionally.

Mathematicians tend to be gregarious, although they often prefer the
company of other mathematicians and are less interested in meeting a wider
variety of people. Naturally, mathematicians occasionally marry other
mathematicians. In comparatively small geographic regions where many
mathematicians live, the potential may exist for too much intermarriage to
lead to a high rate of certain genetic abnormalities. Some concern has been
expressed in the press about this as a possible cause of Asperger syndrome
in communities where its incidence is particularly high, such as Silicon
Valley in California, but the scientific basis for this concern at the epi-
demiological level has not been established.

Imprisonment has sometimes been found conducive to mathematical
research. After Napoleon's retreat from Moscow, Jean Victor Poncelet developed his theory of projective geometry while a prisoner-of-war in Russia,
while the algebraic number theorist André Weil, after being released from
prison in Rouen, said that he might have been able to prove the famous
Riemann conjecture had he been allowed to remain.

Mental health disorders do not seem to affect mathematicians dispro-
portionately, although there have been few attempts to verify this. Perhaps
the first relevant survey was that by Felix Post (1994). This was an investigation into the prevalence of various psychopathologies in almost three hundred outstandingly creative men in science, thought, politics, and art. Only
a few mathematicians were included, namely Charles Babbage, George
Boole, Carl Friedrich Gauss, and William Rowan Hamilton, although there
were also several mathematical physicists. The choice of subjects was determined by the availability of adequate biographies. Family background,
physical health, personality, psychosexuality, and mental health were investigated. All the subjects excelled in their fields by virtue not only of their
abilities and originality, but also their drive, perseverance, industry, and

meticulousness. With a few exceptions, they came across as emotionally warm, with a gift for friendship and sociability. Most had unusual personality characteristics, and minor "neurotic" abnormalities were probably more common than in the general population. The lowest prevalence of psychic abnormalities was in the scientific group of subjects, but even so, nontrivial abnormalities were still found in two-thirds of that group. It is interesting that many of the mathematicians in this book had Asperger syndrome. We know that genes have multiple effects, and we hypothesize that the genes that produce the social deficits of Asperger syndrome are also involved in producing great mathematical ability. Examples in this book are Lagrange, Poincaré, Cauchy, Galois, Riemann, Cantor, Hilbert, Ramanujan, Fisher, Wiener, Dirac, and Gödel.

In 1998, the psychologist Simon Baron-Cohen and his colleagues published the results of a major survey involving undergraduate and graduate students at Cambridge University who were studying either science (physics, engineering, or mathematics—641 students) or literature (English or French —652 students). The students were asked about the incidence of a range of psychiatric conditions in their families, not about their own mental health. Among the relatives of the science students, six cases of autism were reported; among the replies of the literature students, there was just one case. Twice as many cases of bipolar disorder were reported in the families of students studying literature; one hundred cases versus fifty in the families of science students. For other conditions there was not much difference.

There have been other surveys that tell us something about the psychology of mathematicians. For example, Daniel Nettle (2006) conducted one into *schizotypy* and mental health among poets, visual artists, and mathematicians, including twenty-six research mathematicians. Schizotypy scales measure a propensity to loose and unusual thought. Poets, in the survey, scored more highly than the general population, mathematicians less highly. As Nettle observes, this is consistent with a more autistic profile for mathematicians—strong and narrow interests and a high degree of focus. The mathematicians also came out as more introverted and less emotionally labile—this could also be reconciled with the autistic view.

Mathematicians and the Arts

Musical Mathematicians

While there certainly are mathematicians who think of nothing but mathematics, this is unusual. On the contrary, there are plenty of examples of mathematicians who have had other major interests. For example, some took a serious interest in one of the fine arts, such as music, which is often thought to have a special appeal to mathematicians. This may be because musical harmony, according to the Pythagorean theory, is mathematical in nature. The older mathematical dictionaries contain long articles on such theories, as does Archibald (1924), who, in his excellent article on mathematics and music, gives a good account of the earlier literature on the subject. The recent book by Fauvel, Flood, and Wilson (2003), which has excellent illustrations, begins with quotations from a number of people, going back to classical times, who believed there was some relation between mathematics and music.

According to Leeson (2000), the composer Mozart was fascinated by mathematics, but we are not aware of any other composers or performers who were particularly interested in mathematics. However, it is easy enough to name people of mathematical ability who were also musical; for example, the many-sided Olinde Rodrigues had some talent as a composer. Richard Dedekind was an accomplished pianist and cellist who composed a chamber opera. Augustus de Morgan, it was said, excelled as a flautist. Janos Bolyai and A. C. Aitken were exceptionally fine violinists. We can also cite the example of the little-known Hamburg mathematician Johann Georg Busch, at whose home some of the compositions of Carl Philipp Emanuel Bach were performed for the first time. Leopold Kronecker was an accomplished pianist and vocalist. Hermann Grassmann was a pianist and composer; some of his three-part arrangements of Pomeranian folk-songs were published. He was also a good singer and conducted a male voice choir for many years. James Joseph Sylvester believed he had a fine voice and took singing lessons from Charles Gounod. Georg Cantor regretted that he had not followed in the tradition of his mother's family of musicians. Felix Hausdorff was an excellent pianist and occasionally composed songs; he aspired to be a composer rather than a mathematician. Albert Einstein had a passion for music, which he appreciated as a way of experiencing and expressing emotion. He was an enthusiastic (but not very good) violinist; Mozart, Bach, and Schubert were his favorite composers. Photographs of him playing the violin show a different Einstein from the more familiar images. When he was world famous as a physicist, he was reported to have said that music was as important to him as physics: "It is a way for me to be independent of people" (James 2003b); on another occasion he described it as the most important thing in his life.

There seems to have been a tendency for mathematicians to marry into musical families—for example, Richard Courant and Jacques Hadamard did so, and when they entertained at home there was always music. The famous concerts of chamber music held at the home of the amateur mathematician and musician Emile Lemoine exerted a great influence on the musical life of Paris in the latter part of the nineteenth century. In the Claparède-Flournoy inquiry into mathematicians' methods, just over half the respondents listed music as their main intellectual distraction.

However, just after the Second World War, Geza Révész, director of the psychological laboratory in Amsterdam, conducted a survey of 180 mathematicians, 220 physicists, 206 doctors, and 136 writers, who were asked to complete a short questionnaire. This revealed that 24 percent of the mathematicians were completely unmusical, as compared with 16 percent of the physicists, 19 percent of the doctors, and 13 percent of the writers; 44 percent of the mathematicians were unmusical, compared with 33 percent of the physicists, 41 percent of the doctors, and 29 percent of the writers; and 56 percent of the mathematicians were musical, compared with 67 percent of the physicists, 59 percent of the doctors, and 71 percent of the writers; while 9 percent of the mathematicians were very musical, compared with 9 percent of the physicists, 6 percent of the doctors, and 11 percent of the writers. Unfortunately Révész, who specialized in the psychology of musicians, provided no information about his methodology. On the face of it, his results do not support the widespread belief that mathematicians generally are particularly musical.

Literary Mathematicians

A number of mathematicians made noteworthy contributions to literature. Omar Khayyam was celebrated for his poetry as much as his mathematics. In the nineteenth century especially, many scientists enjoyed writing verse from time to time. Augustin Cauchy wrote verse in both French and Latin. Hamilton and J. J. Sylvester fancied themselves poets; in fact the latter believed his poetry would stand comparison with that of Milton. The poems of C. L. Dodgson (Lewis Carroll) are well known, especially *The Hunting of the Snark* and others that occur in the *Alice* books. Perhaps he is the best-known mathematical author, but there were many others. One was Sonya Kovalevskaya, who commented:

> I can work at the same time with literature and mathematics. Many persons that have not studied mathematics confuse it with arithmetic and consider it a dry and arid science. Actually this science requires great fantasy, and one of the first mathematicians of our century [Weierstrass] very correctly said that it is not possible to be a complete mathematician without having the soul of a poet. As far as I am concerned I could never decide whether I had a greater inclination towards mathematics or literature. Just as my mind would tire from purely abstract speculations, I would immediately be drawn to observations about life, about stories; at another time, contrarily, when life would begin to seem uninteresting and insignificant then the incontrovertible laws of science would draw me to them. It may well be that in either of these spheres I would have done much more had I devoted myself to it exclusively, but I nevertheless could never give up either one completely. (Kennedy 1983)

Another mathematician and author was Felix Hausdorff who, as a young man, wrote several works under the pseudonym Paul Mongre, including a play, *The Doctor's Honor,* which ran for a hundred performances on its last run in Berlin. The author of *Men of Mathematics* was also a pioneer writer of science fiction. Hans Freudenthal wrote novels, one of which won a literary prize, under the pseudonym Sirolf.

Mathematicians and the Visual Arts

There is a more obvious connection between the visual arts and mathematics. For example, painters needed to master the laws of perspective, and some have studied these intensively. One was Piero della Francesca, not only a great painter but also one of the outstanding mathematicians of the fifteenth century (see Field 2005). Recently, using computer graphics, students of fractal mathematics have produced beautiful images, some of which look somewhat like mountainous landscapes. Michele Emmer has edited several volumes on mathematics and culture that include articles about the various ways in which mathematics comes into the visual arts (Emmer 1993), and occasionally there are relevant articles in the *Mathematical Intelligencer.*

Mathematical Biography

In the second part of this book we need to rely heavily on the biographical literature. Unfortunately, historians of mathematics are generally accustomed to discussing ideas rather than individuals, so a mathematician's biography and his or her mathematical work are frequently regarded as wholly separate: the life story may provide human interest, but the mathematics is taken to form the heart of the matter. Many of the great mathematicians of the past have been the subject of a full-scale treatment, usually either a *Life and Work* or a *Life and Times,* and there are many more who merit the attention of a biographer.

Biography, in the modern sense, began with the Enlightenment. Before that, biographies were mainly anecdotal in character, when they existed at all. For example, most of what we know about the life of Archimedes of Syracuse comes from Plutarch's profile of the Roman general Marcellus in his *Lives.* The antiquarian William Stukeley collected reminiscences of the great scientist in his *Memoirs of Sir Isaac Newton,* written in 1752 although not published until 1936. The mathematician Augustus de Morgan wrote a biographical essay, published in 1885, entitled *Newton: His Friend: and His Niece.* Although primarily concerned with the personal relations between the niece, Catherine Barton, and the friend, Charles Montagu, Earl of Halifax, de Morgan bravely attempted to fathom Newton's complex personality. Since we can properly claim the inventor of the calculus as a mathematician, de Morgan's essay may be the first biography of a mathematician in something like the modern sense. Of course, numerous biographies of Newton have been published more recently.

Fortunately, a number of mathematical biographies avoid technicalities and emphasize the human side of their subjects; such books appeal to anyone with some knowledge of mathematics and its cultural history. For example, there are the biographies of Augustin Cauchy (Belhoste 1991), William Rowan Hamilton (Hankins 1980), Evariste Galois (Rigatelli 1996), Richard Courant (Reid 1996), David Hilbert (Reid 1970), Jacques Hadamard (Maz'ya and Shaposhnikova 1998), Bertrand Russell (Monk 1996, 1997), George Pólya (Alexanderson 1999), Norbert Wiener (Conway and Siegelman 2005), Ronald Fisher (Box 1978), Oscar Zariski (Parikh 1991), Kurt Gödel (Dawson 1997), Alan Turing (Hodges 1983), and L. E. J. Brouwer (Van Dalen 1997, 2005). Some of these books are more scholarly than others, but the best have been carefully researched, and one should not be put off by fanciful titles. The biographies of Sonya Kovalevskaya by Kennedy (1983) and Koblitz (1987), that of Ramanujan by Kanigel (1991), that of Stefan Banach by Kaluza (1966), and those of Niels Abel and Sophus Lie by Stubhaug (2000, 2002), are also of this type. The authors have some mathematical background, although not all would claim to be professional mathematicians. Kaluza is a journalist, and Stubhaug describes himself as a writer, while Kanigel is an engineer. The life of Nobel laureate John Nash by Sylvia Nasar (1998) is another such example; the author is an economist and now a professor of journalism at Columbia University. As well as books, there are some interesting major articles, such as that on Joseph-Louis Lagrange's personality by George Sarton (1944) and that on Newton's personality by Milo Keynes (1995).

Several collections of mini-biographies of mathematicians are also informative. We have already mentioned Bell's *Men of Mathematics*; there is also James's *Remarkable Mathematicians*. Today, *Men of Mathematics* is rather looked down upon by the historians of mathematics, although it continues to influence the impression that ordinary mathematicians have of the cultural history of their discipline. Bell ranges from ancient times to the end of the First World War, while James begins with Euler and ends with von Neumann. Sketches of the lives of fifteen female mathematicians born between 1776 and 1919 may be found in Case and Leggett (2005), and mini-biographies of fifteen mathematicians associated with the Hilbert problems are contained in Yandell (2002).

Mathematical autobiography seems to be quite a recent phenomenon. The novels of Sonya Kovalevskaya are based to a large extent on her own life. Other important examples are the account of his early life by André Weil (1992) and the memoirs of Norbert Wiener (1953, 1956), P. S. Alexandrov (1979/1980), Mark Kac (1985), and Stanislaus Ulam (1976). The three-volume autobiography of Bertrand Russell (1967–1969) is also interesting, although it is believed he delegated the writing of the third volume to an assistant.

Savant Skills and Other Phenomena

A savant, in this context, is an individual with an islet of outstanding skill in one area, often in the presence of modest or even low general intellectual ability. The skill can appear suddenly, without explanation, and can disappear just as suddenly. It occurs within a narrow range of mental functions. Savant skills may take various forms, which Beate Hermelin (2001) classifies under different headings according to whether they concern mathematics, music, the visual arts, or language. The savants that are of interest to mathematicians are the calculating prodigies, or "lightning calculators," and, to a lesser extent, the calendrical calculators, who can tell on which day of the week a particular date fell. Such people are rare. Between 2 and 3 percent of the general population have some degree of mental disability, but only 0.06 percent of these people are estimated to possess an unusually high level of specific ability that is far above that of the average person. People with these abilities used to be called idiot savants when the contrast with their other abilities was particularly striking, but this term is now obsolete.

Useful reviews of the earlier literature on the subject are S. B. Smith (1983) and Darold Treffert (1989). Scripture (1891) gave an account of some of the outstanding savants of the eighteenth and nineteenth centuries, as did Rouse Ball in his *Mathematical Recreations and Essays* of 1892. One well-known savant was George Parker Bidder, who as a child and a youth gave exhibitions in England and Scotland. He could mentally determine the logarithm of any number to seven or eight decimal places and, apparently intuitively, could obtain the prime factors of quite large numbers. Bidder retained his powers throughout his life and made great use of them in his profession as an engineer; he often tried, without success, to describe the procedures by which he calculated. His son, likewise gifted intellectually, was also a lightning calculator, though not as prodigious. Other lightning calculators whose particular feats are described in the literature are Jedediah Buxton, Zerah Colburn, Johann Martin Zacharias Dase, Henri Mondeux, Vito Mangiamele, Jacques Inaudi, Pericles Diamondi, and Shakuntala Devi. The last-named shows that female lightning calculators do exist, although they seem to be unusual: one estimate puts the male/female ratio at six to one.

In the literature, lightning calculators are usually divided into two groups, referred to as early/auditory and late/visual, depending on when they started to develop their exceptional abilities and how they mentally represent numbers. Auditory calculators "hear" the numbers in their heads when calculating, and their calculation is often associated with some verbalization or exaggerated motor activity. Their abilities evolve at an early age, long before they learn written numeration, and they are therefore sometimes called early or natural calculators. By contrast, visual calculators "see" the numbers mentally and stay relatively quiet when calculating. They usually learn to calculate after learning to read and write; some of them start to develop their abilities only during their teenage years and are therefore called late calculators.

The spontaneous pleasure savants take in their skill is palpable, as is the facility with which it is mastered. They are usually quite indifferent to how these activities strike others. People with limited mental capacities, as well as a tendency toward repetitive behavior, as we find in those with autism, may concentrate more on their natural special abilities than do those who have more choices. One consequence of such preoccupation is that continuous repetition of subject matter might help the natural emergence of knowledge about its structural properties, though such knowledge may remain unconscious. This is what seems to happen with calendrical calculators. The neurologist Oliver Sacks (1995), observing a famous pair of savant twins, saw them playing a game where they rapidly alternated calling out to each other increasingly large prime numbers of up to twenty digits. Sacks said that the twins could recall numbers up to three hundred digits; such an outstanding memory for numbers has been reported in nearly every case of savant mental calculation, although such remarkable memories are not confined to savants. According to Radford (1990), savant memory seems to be more automatic and literal than that of a prodigy of any sort. When the feat is simply one of memory rather than calculation, the term "savant" seems less appropriate.

Savant ability is frequently found in people with some form of autism. It is estimated that at least one in ten of those with an autistic spectrum disorder displays some specific skill at a high level, and a high proportion of savants are considered to have suffered from some form of autism. Hermelin (2001) describes one 20-year-old with autism, the son of university graduates in mathematics, who showed no responsiveness to people and seemed not to understand language at all, although his visuospatial intelligence was very high. This young man, without any training, was able to identify prime numbers and factorize composite numbers much better than any person with normal intelligence.

It also seems that training can help to develop savant skills. A recent case of a self-trained calculating prodigy has been investigated by the research group of Mauro Pesenti in the Belgian city of Louvain. The subject was a young insurance broker who, around the age of 20, trained himself to achieve a high level of expertise after becoming aware that he had exceptional abilities. The tests he was given showed that the most important characteristic of his abilities was the high efficiency of his long-term memory storage and retrieval processes. Along with a very good short-term memory capacity, his intuition and knowledge of calculation algorithms made complex calculations possible, while training automated the sequence of steps in the algorithms, thus decreasing the short-term memory loss in complex calculations. Details are given in Pesenti (2005).

Savant arithmetical calculators often show a fascination for numbers at a very early age, counting things continually; consequently they become familiar with inter-numerical relationships, although these become evident only when such an individual's increasing skills at calculation are noticed. There are marked individual differences in the strategies employed by great mental calculators without any cognitive impairments. Some use shortcuts and near guesses; others use detailed rapid calculation. All are able to deal instantly with numerical properties of any numbers they come across. Some have an outstanding numerical memory; others not.

Those having the talent are impatient with scientists who try to probe their abilities. Hermelin (2001) quotes one savant who said with exasperation, "I just do it. Just as it seems natural to you to formulate a sentence without consulting the rules of grammar or tallying the meaning of each word, I calculate." Their attempts to explain their gifts are unhelpful, even baffling, and their methods mysterious. Another savant said:

> If I was asked any question, rather a difficult one by itself, the result immediately proceeded from my sensibility without my knowing at the first moment how I had obtained it; starting from the result, I then sought the way to be followed for this purpose. That intuitive conception which, curiously enough, has never been shaken by an error, has developed more and more as needs increased. Even now, I have often the sensation of someone beside me whispering the right way to find the desired result; it concerns some ways where few people

have entered before me and which I should certainly not have found if I had sought them by myself. It often seems to me, especially when I am alone, that I find myself in another world. Ideas of numbers seem to live. Suddenly, questions of any kind rose before my eyes with their answers.

Mitchell (1907), after reviewing many examples from the eighteenth and nineteenth centuries, concluded that "skill in mental calculation is . . . independent of general education; the mathematical prodigy may be illiterate or even densely stupid, or he may be an all-round prodigy, and veritable genius." Some professional mathematicians have possessed savant skills, although this is unusual. François Arago said of Leonhard Euler that he calculated without apparent effort, as men breathe or as eagles sustain themselves in the air. The skill was also possessed by John Wallis, Gauss, Hamilton, Poincaré, Ramanujan, and Banach, among others.

One of the most remarkable and best-documented lightning calculators of the twentieth century was the New Zealander A. C. Aitken, who cooperated with the psychologist A. M. Hunter in a scientific investigation of his abilities (Hunter 1968, 1977). When Aitken said of a number "it feels prime," it invariably was. This gift may have had a genetic component, since his father's elder brother also possessed formidable skills in mental arithmetic. "The secret, to my mind, is relaxation," Aitken once wrote, "the complete antithesis of concentration as usually understood." Aitken was neither a primarily visual nor auditory calculator—rather, he was a combination of both and something more enigmatic: "I myself can visualise if I wish at intervals in a calculation, and also at the end, when all is done, the numbers come into focus, but mostly it is if they were hidden under some medium" (Hunter 1968).

On another occasion Aitken observed:

> I have noticed at times that the mind has anticipated the will. I have had an answer before I even wished to do the calculation. I have checked it and am always surprised that it is correct. This I suppose (but the terminology may not be right) is the subconscious in action; I think it can be in action at several levels; and I believe that each of these levels has its own velocity, different from that of our ordinary waking time, in which our processes of thought are rather tardy. (Hunter 1968)

Aitken also had astonishing memorization abilities, though these are not always associated with calculating skills. His memory was photographic; he would scan a text rapidly and it would be permanently fixed in his mind. Once he recited to the students in his class the first two thousand digits of pi. He occasionally gave public demonstrations. He stressed the artlessness of his talent, saying one didn't force the lock of discovery, one simply allowed it to open:

> I believe we are surrounded the whole time by marvellous powers, are immersed in them, closer than breathing, and I think that all great music, poetry, mathematics and real religion come from a world not distant but right in the midst of everything, permeating it. When I wish to do a feat of memory or calculation, or, sometimes, new mathematical discovery, I let slip some sort of cog and lie back in this world I speak of, not concentrating, but waiting in complete confidence for the thing desired to flow in. (Hunter 1968)

Aitken thought of his mental calculation as a "compound faculty" that he was unable to describe adequately.

Cognitive scientists are fascinated by savants, whose inability to provide an account of the processes involved does not necessarily entail a nonexistence of the relevant cognitive mental representations in an abstract form. A paper by the Australian cognitive scientists A. W. Snyder and J. D. Mitchell (1999) includes the following summary:

> Unlike the ability to acquire our native language, we struggle to learn multiplication and division. It may then come as a surprise that the mental machinery for performing lightning-fast integer arithmetic calculations could be within us all even though it cannot be readily accessed, nor do we have any idea of its primary function. We are led to this provocative hypothesis by analysing the extraordinary skills of autistic savants. In our view such individuals have privileged access to lower levels of information not normally available through introspection.

According to Treffert (1988, 1989), there were just twenty-five documented calculating savants in the world at the time that he wrote. One was a 26-year-old Englishman who suffered an epileptic fit at the age of three. Ever since then he has been able to see numbers as shapes, colors, and textures. He not only has savant skills; he can describe what he sees in his head. "When I multiply numbers together," he says, "I see two shapes. The image starts to change and evolve, and a third shape emerges. That's the answer. It's mental imagery. It's like maths without having to think." Recently the young man broke the British and European record by recalling the ratio pi to 22,514 places of decimals (the world record is a lot more). To him pi isn't an abstract set of digits; it's a visual story, a film projected in front of his eyes. He learned the number forward and backward and spent five hours recalling it in front of an adjudicator.

Scans of the brains of autistic savants suggest that the right hemisphere might be compensating for damage to the left. Mathematics and memory are primarily right-hemisphere skills. Frith (2003) quotes the Australian researcher Ted Nettleback, who suggests that savant skills demonstrate the brain's capacity to develop new cognitive modules in certain domains that create a hotline to a knowledge store in long-term memory and operate independently of basic information capacity. Hermelin (2001) also emphasizes that weak central coherence, focusing on detail, and a quasi-independent modular style of thinking give autistic savants an uneven, fragmented cognitive profile.

Hermelin and O'Connor (1990) have put forward another explanation of savant talent. They observe that a specific talent for mathematics tends to be self-contained but that it also includes its own aspect of central processing. We would thus regard savant ability as quasi-modular—circumscribed by, but not restricted to, sensory mathematical processing. In this view, the modules of the mind (in this case mathematical) are self-contained and do not communicate with each other but with a central processing area in the brain. Research is in progress to find out if this is in fact the case.

Snyder and Mitchell (1999) make an intriguing observation:

> Some autistic savants have the ability to keep time for extended periods with accuracy to the second. Apparently our internal clocks are more precise than might have been imagined. For example, when one autistic child was awakened he said it is 2.14 a.m. and went back to sleep. This ability to equi-partition time could also contribute to the impressive musical skills of many autistic savants and, when coupled with equi-partitioning in space, could suggest a mechanism which interrelates music and mathematics.

There is another phenomenon that may have something to do with savant skills. The behavioral scientist Francis Galton (1883/1951) described visuo-spatial number representations he collected from eighty male and female subjects, who were chosen because of the way in which they mentally saw numbers. Galton distinguished these representations from the capacity to visualize mentally in an effortful way a number or an arithmetic operation as it would appear if actually written. The kind of number representation Galton's subjects experienced was more complex and was automatically activated when they saw, heard, or thought of a number. Some number representations concerned simple digits or numbers. For example, some subjects reported seeing numbers as they appeared on dominoes or on playing cards; others declared they saw digits in a specific color. Some subjects reported seeing each number at a given place in a stable spatial mental structure. When these subjects thought of the series of numbers, "They show themselves in a definite pattern that always occupies an identical position in their field of view with respect to the direction in which they are looking." These visual patterns, called by Galton "number forms," either consisted of a simple line, with or without shifts of orientation, or took more complex forms such as rows or grids. They could be colored, present changes in luminosity in some locations, or occupy different planes. Other investigators reported similar data, but added that some subjects also had visuospatial representations for the months of the year or the days of the week.

Number forms varied with respect to their structure, but some common characteristics could be noted. First, there seemed to be a clear intra-subject consistency. For a given subject, the whole number form always had an identical structure, each number always occupying the same position on the form. Similarly, subjects having colored number images always saw the same number associated with the same color. A second common characteristic was the emergence of the number forms during subjects' infancy, and a third was the involuntary and automatic aspect of its activation: the subjects indicated that the forms were automatically activated by any number that was heard, was seen, or came to mind. A fourth characteristic was the subdivision of the number form into a clear and a less precise part. Although he did not carry out a systematic statistical analysis on the incidence of the ability to see number forms, Galton estimated that it might be possessed by about one out of thirty adult males and one out of fifteen females. The phenomenon is thus not exceptional. There appeared to be no regular correlation between seeing number forms and level of expertise in arithmetic.

In 1992 the Louvain research group published the results of some more research in this area, which confirmed Galton's conclusions. A short questionnaire was given to psychology students (153 females, 41 males); those who claimed to see number forms were given a more systematic extended questionnaire. Two spontaneously written introspective reports were collected through the informal inquiry, while the short questionnaire yielded twenty-seven positive answers. Out of the total of twenty-nine subjects, twenty-six answered the extended questionnaire. Each of the twenty-six was questioned at intervals, and their number forms remained much the same. Although additional information about the phenomenon was discovered through this research, the broad conclusion was that it was genuine and not uncommon. Similar investigations in other countries agreed with this. Another inquiry at Louvain involving 217 students from a broader selection of departments showed that forty-two of them reported seeing number forms.

Fias and Fischer (2005) describe more recent research in this area. About 15 percent of normal adults report visuospatial representations of numbers: more systematic studies have supported these anecdotal reports by demonstrating a tight correlation between mathematical and visuospatial skill. In the clinical field, learning disorders indicate a similar association between visuospatial and mathematical disabilities. Evidence from brain imaging provides further support for a link between numbers and space. Sacks (1995) has observed that savant gifts appear fully fledged from the start and have a more autonomous—even automatic—quality than normal talents. He believes that the great lightning calculators often perceive numbers as a spatially extended domain, with an amazing wealth of detail. In the mind of the calculating prodigy, he writes, each number lights up not as a point on a line, but rather as an arithmetical web with links in every direction. Savants summon up—they dwell among—strange scenes of numbers; they wander freely in great landscapes of numbers. They create dramaturgically a whole world made of numbers. They have, Sacks continues, a most singular imagination—not the least of its singularity is that it can imagine only numbers. They do not seem to operate with numbers non-iconically, like a calculator; they see them directly, as a vast natural scene.

Galton also drew attention to the phenomenon known today as *synesthesia,* which occurs when stimulation of one sensory modality automatically triggers a perception in a second modality, in the absence of any direct stimulation to this second modality. He described individuals who saw each number as having a particular color, just as some people who have absolute pitch in music associate a particular color with each note. Other people, notably Poincaré, have seen the letters of the alphabet in color. These phenomena, which are fairly common, have been much studied but remain something of a mystery. The sensations are automatic and cannot be turned on and off. Women are more likely to have synesthesia than men. People are generally born with it, and it runs in families. For some further information see *Phantoms in the Brain,* by Ramachandran and Blakeslee (1998).

Chapter 2 } Mathematical Ability

The Making of a Mathematician

In his best-known book, *Hereditary Genius* (1869), Francis Galton presents some evidence in favor of the inheritance of scientific ability but then concludes: "It is I believe owing to the favourable conditions of their early training that an unusually large proportion of the sons of the most gifted men of science become distinguished in the same career." Notwithstanding historical stress on the biological basis of exceptional ability, the value of a stimulating environment in the home and at school in facilitating development and academic achievement is well established. There exists a large body of empirical evidence testifying to the value of early stimulation and encouragement to learn. Many young people find it hard to do effortful things on their own, especially when these require sustained concentration, with study and practicing being about the least favored activities for most youngsters, even the highly able.

The early experiences of eminent mathematicians do not seem very different from those of other eminent scientists. The psychologist Anne Roe (1951), in her study of eminent scientists in America, concluded that the following seemed important for the development of a first-rate research scientist: a background in which intellectual activities are valued, an intense emotional need that can be satisfied by creative activities, and the ability to concentrate energies on research (probably reinforced by early experience of rewards for such concentration). She found cases in which at school the student was amazed to find that the branch of science being studied was not already fully researched. Motivated by the desire to contribute something new, the student then began to study that particular subject obsessively.

These conclusions were later confirmed by the applied psychologist William Fowler (1983) and his research associates, who studied the early experiences of twenty-five outstanding mathematicians of the past. Fowler concluded that certain factors served as developmental prerequisites but that these factors assumed various forms. Most of the prerequisites, but not necessarily all of them, needed to be present in each case. One factor was experience of special cognitive stimulation in early childhood, at least by the age of 5. It is intellectual stimulation that matters, not so much the acquisition of mathematical skill and knowledge; although verbal concepts were invariably stressed much more than mathematics, these served to establish the potential to acquire mathematical skills. Another factor was the presence of intellectual models from earliest childhood, and a third was the successful generation of intense self-directed study during the early years. The early lives of some of these mathematicians were marked by emotional deprivation, such as the loss of a parent by death. A serious illness in childhood occurred in several cases.

Although Fowler's choice of subjects only includes twenty-five mathematicians, his conclusions seem to remain valid when tested against the sixty who are profiled in *Remarkable Mathematicians*. These mathematicians, who were not exclusively European, were mainly born in the eighteenth and nineteenth centuries. The majority were of humble origins, although some came from quite well-to-do bourgeois, or even wealthy, families. Some displayed an enthusiasm and aptitude for mathematics from an early age, but others decided to become mathematicians relatively late. Intellectual exchange with mentors was important in some cases—sometimes a school-friend, more often a perceptive schoolteacher. It is well-established (see Sulloway 1996) that firstborn children, especially firstborn sons, have an advantage over their siblings, because they are given first call on the resources of the family.

In most cases, we can identify at least one educated close family member who served as an intellectual model for the young mathematician. For example, Leibniz's father was a professor of moral philosophy at Leipzig, and his mother also came from an academic family. Fermat's father was a lawyer, and his mother came from a family of magistrates. Pascal's father was a lawyer with an interest in mathematics; Maria Agnesi's father, a professor of mathematics at the University of Bologna, encouraged his eldest daughter to pursue scientific interests by hiring various distinguished professors to tutor her. Sophie Germain's father, a deputy of the estates general and director of the Bank of France, owned an extensive library to which she had access. The mathematicians in the quarrelsome Bernoulli family learned from each other. Poncelet's foster family possessed a complete library of French classics, all of which he had read by the age of ten. Sonya Kovalevskaya's father was a well-educated man who spoke English and French fluently; Emmy Noether grew up in a home constantly visited by members of a stimulating circle of scholars. Norbert Wiener's father was an unusually well-read professor of languages at Harvard.

Galton mentions the common belief that the mothers of great men are more influential on their development than the fathers. Certainly Carl Friedrich Gauss, Henri Poincaré, and David Hilbert believed that their mathematical gifts came from their mothers' rather than their fathers' sides of the family. Regrettably, far too little is known about the mothers of the great mathematicians, although some of them were clearly remarkable women. Often we just know their married and maiden names; sometimes not even the latter. Ada Byron's mother, a mean and selfish hypochondriac, was interested in mathematics herself. The mother of Evariste Galois was described as "intelligent, lively, headstrong, generous, and eccentric," Henry Smith's as a "woman of learning and ability." Sonya Kovalevskaya's mother was a sociable woman and an accomplished pianist whose antecedents were of some intellectual distinction. The mother of Ramanujan was described as "shrewd, cultured, and deeply religious," and Andrei Nikolaevitch Kolmogorov's mother as "liberal and independent."

Until fairly recently it was not uncommon for children to begin their education at home rather than at school. The most frequent instructor was the father, or both parents; in a few cases it was the mother alone. An uncle or an aunt was often an important mentor. Probably this kind of education was similar to what the children might have received at school, but they experienced it at an unusually early age. Once they had acquired the basic skills of reading, writing, and arithmetic, they might be taught some Latin, but they were unlikely to have had any early exposure to mathematics. Where parents themselves lacked financial resources and social opportunities, their child's abilities might attract the support of wealthy patrons or public aid—for example, the deputy head of the École du Génie in the case of Gaspard Monge, some citizens of Auxerre in the case of Joseph Fourier, the Duke of Brunswick in the case of Carl Friedrich Gauss, August Leopold Crelle in the case of Niels Abel, General Foy in the case of Lejeune Dirichlet,

Antonin Dubost in the case of Elie Cartan, and an "intellectual of French origin" in the case of Stefan Banach. Details may be found in James (2002).

Fowler's work has been taken further by the psychologist Michael Howe (1999), who warns that while there is much to be said for early intellectual stimulation, there are also likely to be costs that may keep the child from making the best of his or her abilities later in life. Children whose early years are filled with learning sessions may have insufficient time to play with other children and to learn to make friends. A child whose activities are largely determined by a parent's bidding may be overly conscious of the necessity to please other people and may fail to develop personal enthusiasms and interests. Young people who are deprived of opportunities to make choices and decisions for themselves, to choose their own friends and make their own mistakes, do not learn to look after themselves and survive life outside home and away from the family. They may enter adulthood without a sense of purpose and with none of the confidence likely to be found in those who have been encouraged to take on responsibilities and acquire the various qualities and skills that make people independent, self-aware, and confident enough to pursue their own goals.

Recently, neuroscientific research into the development of the brain during the first few years of life has begun to play a role in education policy debates. In early childhood, the brain changes enormously and undergoes several waves of reorganization. It is not the neurons themselves that change but the wiring—the intricate network of connections between them. Some childcare centers in the United States have begun to structure their curricula around the idea that children should begin the study of mathematics, languages, logic, and music as early as possible and not have to wait for formal schooling to commence at the age of five or more. However, neuroscience does not provide a firm basis for this idea at the present time.

Mathematical school education is another subject with a vast literature, much of it American. In terms of mathematical achievement, American schoolchildren compare poorly with those in other countries. Various explanations have been offered, but the main difference seems to be simply that they spend less time on mathematics and that the time they do spend is used inefficiently. Societal attitudes tend to magnify the basic differences. American parents tend to value literacy more than mathematical skills. Also, they believe that skill in mathematics is innate, and that ordinary children cannot be expected to learn what they are asked to learn. Yet

mathematically gifted American children excel in the International Mathematical Olympiad competitions, at which young mathematicians from all
over the world compete against each other in problem-solving.

In the Soviet Union special schools were established where mathematically gifted children could progress rapidly. Kolmogorov, who taught at one
of these schools, thought that the earlier the stage of general human development at which a person stops, the higher his or her mathematical
talents (Arnold 1993). Russian educational psychologists observed that such
children showed a profoundly emotional regard for mathematical work.
Details may be found in Krutetskii (1976). Krutetskii writes that it used to be
believed that "gifted children were inferior to ordinary, normal children in
every respect except intelligence. Gifted children were alleged to be physically weak, sickly, unattractive, emotionally unstable, and neurotically inclined." Galton collected evidence to refute this belief, and subsequent study
by psychologists led to the establishment of what was in almost every way
the opposite picture. Yet there seems to be some truth in the idea that gifted
children tend to have certain distinctive biological attributes, as shown by
Benbow (1988a).

Myopia has frequently been associated with higher general intelligence.
The condition, in which the eyeball is too long from front to back to
function normally, affects the personality as well as the eyesight (see Young
1967; Trevor-Roper 1988), and there is certainly a genetic factor involved.
Among the great mathematicians of the past, Sophus Lie, Henri Poincaré,
Tullio Levi-Civita, Emmy Noether, Ronald Fisher, and Norbert Wiener
were strongly myopic. Benbow found that among gifted American schoolchildren, myopia is four times as prevalent as among schoolchildren generally. The reason for this is not known.

Symptomatic atopic disease (allergies) is also associated with high ability.
Benbow found that extremely precocious mathematical or verbal reasoners
are about twice as likely to have allergies as members of the general population. Such extremely able students were reported by their parents to have a
higher incidence of allergies than their parents, their siblings, or a comparison group of students.

Throughout history left-handed people have been regarded as inferior to
right-handed people in many ways, and this prejudice lingers today. Although no overall differences in general intellectual functioning between
left- and right-handers have been found, there may be differences in certain

aptitudes (McManus 2002). Recently, left-handers have been found to show high performance levels in tasks mediated by the right hemisphere of the brain. The right hemisphere is specialized for nonverbal tasks and the left for verbal, although these differences are quantitative rather than absolute. Non-right-handedness has been positively associated with spatial ability and musical ability. Relatively high frequencies of left-handedness have been found among university mathematics teachers and students, as well as dyslexics.

Mathematical reasoning ability, as opposed to computational ability, is believed to be more strongly under the influence of the right hemisphere. Indeed, left-handed schoolchildren have been found to have a slight advantage in numerical reasoning. Furthermore, left-handedness and mixed-handedness have been related to extreme mathematical talent. Compared to a control group, American students who scored extremely well (the top one in ten thousand students) on a test of mathematical reasoning ability were much more likely to be left-handed.

A more extensive attempt to identify and encourage mathematical talent was the American Study of Mathematically Precocious Youth (SMPY), initiated in 1971 by Julian C. Stanley at the Johns Hopkins University. Similar projects were started elsewhere, and eventually some seventy thousand students were included each year. The SMPY sought first to identify mathematically gifted children, essentially by standard school tests of achievement, and then went on to apply a policy of selective acceleration that enabled students to pursue each ability as far and as fast as they were able and wished to.

Benbow (1987, 1988b) reported that questionnaire responses from 1,896 SMPY students showed them, as compared to control groups, to be superior in both ability and achievement, to express stronger interest in mathematics and science, and to be more strongly motivated educationally. The majority felt that SMPY had not detracted from their social and emotional development: they experienced increased zest for learning and for life, better attitudes toward education and other activities, enhanced feelings of self-worth and accomplishment, and a reduction of egotism and arrogance.

Benbow was particularly interested in American high school students with extreme mathematical talents. She concluded that they are typically male, left-handed, and myopic, and have a higher than average incidence of allergy, migraine, and immune disorders. She proposed that this collection of physiological correlates suggests enhanced development of the right hemisphere of the brain and is a biological prerequisite for exceptional mathematical ability. The logic is that heightened development of this side of the brain enhances visuospatial capacity, a known predictor of superior mathematical reasoning ability. More recent research indicates that both hemispheres are important, and there is growing evidence that the brains of the mathematically gifted may be functionally organized differently than those of people with average mathematical ability. The two halves of the brain cooperate better, and this is what underlies giftedness, not just in mathematics but in music and the arts in general.

When attempts have been made to compare children's mathematical competence across societies, dramatic and consistent differences have been revealed. Every four years, researchers at Boston College issue a report called *Trends in International Mathematics and Science Study* (TIMSS), which covers forty-nine countries. The most recent of these large-scale studies showed typical results. Students from all participating Asian countries (Singapore, South Korea, Japan, and Hong Kong, among others) showed the highest average performance. North American students compared poorly with these students.

Is There a Gender Difference in Mathematical Ability?

Do women have less aptitude for mathematics, or for science generally? There is no evidence to suggest that scientific ability is wholly genetic, or that men are more likely to have the relevant genes than women. There is evidence, however, that the brains of men and women are subtly different, that they function differently, and that, at the psychological level, males and females are interested in different aspects of their environment. In Cambridge, Simon Baron-Cohen (2003) and his colleagues have been investigating this using personality questionnaires. The results suggest that women as a group score higher than men for empathy (i.e., interest in other people and their emotional lives), while men score higher than women for interest in systems of different kinds, such as diagrams and machines.

At one time there was a widespread belief that women were incapable of abstract thought. Once this was shown to be quite untrue, women were warned that too much study would endanger their health. They encountered various social and formal barriers to their intellectual development, particularly in the sciences. Women were not allowed to enroll at universities in France until 1861, nor in Germany until 1900. Before that, university professors frequently refused permission for women even to attend their lectures. At Cambridge University women were allowed to sit for examinations but were not permitted to take degrees until 1948. Discrimination extended to the scientific academies. The Royal Society of London elected no women until 1945; the Paris Academy of Sciences not until 1979. In the United States colleges and universities had "anti-nepotism" rules that prevented two members of a family from holding academic positions in the same department. When a married couple wrote a book, usually the husband was assumed to be the principal author. Of course, such prejudices are not peculiar to mathematics.

The only female mathematician to appear in *Men of Mathematics* is Sonya Kovalevskaya, who is included in the profile of Karl Weierstrass. In *Remarkable Mathematicians,* only three of the sixty subjects are women. There can be no doubt that many other women were gifted mathematically but lacked the opportunity to develop their gifts. For her excellent book *Women in Mathematics,* Claudia Henrion (1997) interviewed nine successful female American mathematicians in academia. She was mainly interested in the difficulties they encountered in their careers and the ways they dealt with them. She found it useful to distinguish between expectations about women held by society at large and those held by the mathematics community in particular. As she explains, for a mathematician to be "one of the boys" is often subtly helpful in career terms, and this can pose problems for female mathematicians. In North America, at least, mathematics is still too much of a man's world, as can be seen from the articles in Case and Leggett's (2005) book on the subject. Lynn Osen's book *Women in Mathematics* of 1974 describes the careers of another seven female mathematicians. An American study, conducted at the Institute of Personality Assessment and Research by Ravenna Helson (1980), examined a sample of creative female mathematicians and found them to be individualistic, original, preoccupied, artistic, complicated, courageous, imaginative, and self-centered.

The idea that women are incapable of abstract thought died long ago, but the idea that there are some differences between the sexes in mathematical ability has been studied over and over again. Benbow (1988b) is a valuable survey of earlier research, while educational psychologist Jeff Evans gives a useful review of recent research in *Adults' Mathematical Thinking and Emotions* (2000), one of a series of monographs by different authors on aspects of mathematical education. Evans writes that there are two main schools of thought among educational psychologists, the biological and the cultural. The cognitive scientist Brian Butterworth, author of *The Mathematical Brain* (1999), believes that cultural pressures and variations in the quality of mathematics teaching are likely responsible for what he calls "one of the most enduring mathematical myths." Decades ago, girls tended to be taught mathematics by nonspecialist teachers, unlike boys. As a consequence of their better education, the boys did better, but today girls are catching up, and there is evidence that they have yet to reach their full potential.

Maccoby and Jacklin (1974) and Anastasi (1958) reported sex differences in various areas, including mathematical ability. In mental capacity and intellectual interests boys and girls have much in common, the range of difference in either sex being far greater than the difference between the sexes. In England girls easily outperform boys in all subjects at all ages, with one exception—mathematics, where girls still outperform boys but only by a narrow margin. Geary (1996) reviewed a wide range of industrialized countries to show that boys, on average, outperform girls in solving mathematical problems, and the results of the International Mathematical Olympiad tend to confirm this. Among American teenagers more boys than girls score in the upper reaches on the Scholastic Aptitude Test for mathematics. The difference between boys and girls increases with age, but at seventeen years the average difference between boys and girls is only 1 percent. The most recent TIMSS report shows little scholastic difference between the sexes, except for science among 14-year-olds, in which boys did better than girls in most countries. Cross-national comparisons using the same tests in each country reinforce the overall picture that in most countries, including the United States, there is no statistical difference in ability between boys and girls. However, although this is well-established at normal levels of mathematical ability, it seems there is some gender difference at extreme levels. The Johns Hopkins SMPY study found that boys showed more frequent and more extreme giftedness in mathematics.

Further investigations more specifically focused on different kinds of mathematical ability has been collected and reviewed by Stage et al. (1985), who concluded that in American high schools boys perform a little better than girls on tests of mathematical reasoning, while boys and girls perform similarly on tests of algebra and basic mathematical knowledge. Also, girls occasionally outperform boys on tests of computational skills. More recently the available data was analyzed by Hyde et al. (1990), who concluded that any differences vary with age and the particular type of ability being assessed. Overall the sex difference is negligible, but in the direction favoring females. Tests of computational skill show a small gender difference favoring girls in childhood, but not in older girls. Among college students, males do slightly better on tests of problem solving. There is no gender difference in the understanding of mathematical concepts at any age.

In the United States, popular belief holds that, on average, boys perform better than girls in mathematics at school and university. This has led to reduced expectations of good mathematics results by parents of girls and a reduced interest from the girls in mathematics, engineering, and science. In Britain, on the other hand, there is no evidence of any differential encouragement by parents for the study of mathematics; moreover, British girls routinely outperform boys in school-leaving examinations, including those used to assess ability in mathematics and science. In Norway a survey of 356 sixth-grade and 353 ninth-grade students at several schools by Skaalvic and Rankin (1994) showed no sex difference in mathematical achievement. Boys had higher mathematical self-concept, self-perceived mathematical skill, and mathematical motivation than girls, but lower verbal motivation. Girls showed higher levels of verbal achievement than boys, although there was no significant difference in verbal self-concept or self-perceived verbal skills. Some investigators have wondered whether girls tend to do better in mathematical tasks in which it is easier to use verbal strategies. Where verbal strategies are arguably less useful, for example geometry, probability, and statistics, girls score lower than boys.

Mathematical Prodigies

An extensive literature exists on the subject of prodigies. In *Nature's Gambit* (1986), D. H. Feldman lists ten characteristics of people with these extraordinary abilities. First there is family history: when you find a prodigy you also tend to find a family history of interest in the field in which the prodigy emerges. Next, the exceptional talents tend to be focused in a specific area. Then sufficient resources need to be available to sustain the process of talent development. Prodigies are often first-born males. Prodigies are single-minded: they display a passion for their chosen field combined with an ambition that brooks no compromise. They also possess an inner confidence—a sense that they carry a special ability—and when they encounter the field that allows this special ability to express itself, a rapid engagement with and passionate embracing of that field usually follow. Most prodigies experience a crisis in their lives at some time during adolescence. This crisis seems to be a function of changes both internal, having to do with their understanding of their own mental processes, and external, having to do with the realization that they must soon make their way as adults. Prodigies often arouse hostility or at least ambivalence. They tend to display a mixture of childlike and adult qualities: the focus on developing their talents is often accompanied by an overprotective environment, so prodigies may not make the transition to adulthood as completely as most others do.

From time to time child prodigies emerge in mathematics, as they do in music and in languages. Their parents may report noticing something exceptional in their child early on, such as a passion for counting, a liking for numbers (often including them in stories and rhymes), a use of logical connectives (such as "if," "then," "so," "because," "either," "or"), a delight in making patterns (showing balance and symmetry), neatness and precision in arranging toys, the use of sophisticated criteria in sorting and classifying, and a pleasure in jigsaws and constructional toys. They learn much faster than normal children and represent information in their domain in an atypical manner. Mathematical child prodigies solve problems in idiosyncratic, apparently intuitive ways, and they may thus have difficulty when forced to shift to conventional methods. Unlike savants, prodigies are capable of creative achievements, given the right kind of upbringing and the opportunity to show what they can do. All too often, however, the prodigy has difficulty being accepted by the community and never has an opportunity to make fruitful use of his or her gifts. Certainly there is no guarantee that a child prodigy will develop the qualities that exceptional mature accomplishments depend on, although early promise often carries through to later achievement. Some suffer from developmental disorders that give them extraordinary mathematical talent on the one hand but a complete inability to handle their daily lives on the other.

It is easy to find examples of mathematicians who enjoyed a stimulating and supportive early upbringing, were prodigies in childhood, and as adults were capable of creative achievements. The Italian Maria Agnesi was one of these, as were the French Blaise Pascal and Evariste Galois, the Irish William Rowan Hamilton, and the Hungarian John von Neumann and Paul Erdös. They seem to fit the characteristics listed by Feldman fairly well, except that in some cases we do not know of a family history of interest in mathematics.

We can also find examples of mathematical child prodigies who experienced a stimulating and supportive early upbringing but were not capable of creative achievements as adults. There are various reasons why this might be so. For example, a mathematician may be exceptionally well equipped with mental skills and knowledge but not have the drive necessary to fuel the lengthy and concentrated intellectual efforts required to overcome difficulties and obstacles.

We do not need to look far to discover a pair of mathematical child prodigies with much in common, one of whom was outstandingly successful, the other a complete failure. It seems highly likely that both suffered from Asperger syndrome, as did their fathers. The name of Norbert Wiener is well known today, but the name of William James Sidis has been largely forgotten. Their respective fathers believed that genius was the result of nurture rather than nature.

Boris Sidis, the father of William, was born in the Ukraine in 1867. As a child he knew several languages, was well read in history, and composed poetry. He emigrated to the United States before the age of 20 and, after some months working in the sweatshops of New York, settled in Boston, Massachusetts. There he met his future wife, Sarah Mandelstaum, who was also a Ukrainian immigrant. She was studying at Harvard; later she transferred to the Boston University School of Medicine. Within a year they were married. Boris was admitted to Harvard, where he became a protégé of the philosopher and pioneer psychologist William James. In the course of his successful career in psychotherapy he wrote seventeen books and fifty-two articles.

The prodigy William James (Billy) Sidis was born to Boris and Sarah on 1 April 1898 (we refer to the biography by Amy Wallace [1986] for the facts about his life). They also had a daughter, of whom Boris remarked, "If she had not been born a woman she would have been very intelligent." Later Billy was to express over and over again his hatred of his mother, perhaps because she did not stand up to his father. A precocious child to a high degree, Billy never had any exposure to outside games or other childish activities. He was years behind the development of normal children in activities such as grooming, tying his shoelaces, or dressing appropriately. He consumed his food in a peculiar way. While still a child he wrote at least four books and invented a kind of Esperanto. He liked taking examinations; by the age of seven he had passed the anatomy examination for Harvard Medical School and the entrance examination for the Massachusetts Institute of Technology (MIT).

Billy had a phenomenal memory that enabled him to learn foreign languages rapidly. He fits the profile known to psychologists as the "little professor," overflowing with information and anxious to impart it to anyone who came near. Mathematics, in which he displayed savant skills, was his favorite subject. In 1910, at the age of 12, he gave a two-hour discourse at the Harvard Mathematical Club entitled "Four-Dimensional Bodies." According to Wiener, his talk would have done credit to a graduate student. After it, the press described him as the world's greatest child prodigy. A photograph of him shows a lad with a wary, riveting gaze, with something strange and intense in his gray eyes. His voice was somewhat soporific, and he laughed at his own jokes, which weren't funny at all. He had a quick temper, which made him keep away from others. He was shy of strangers, and it was hard to make him change his opinions.

Before being admitted to Harvard, Billy was educated at home, using Montessori methods. Once at university he was treated cruelly by his classmates. A misfit with no social instincts, he was horribly ostracized and made the butt of practical jokes. Fascinated by bus and train timetables, he produced a magazine on the subject, and later wrote a book called *Notes on the Collection of Transfers*. He always dressed in the same peculiar way and, like Ramanujan, was said to be careless about personal hygiene. He developed psoriasis; his sensitive skin made it difficult for him to shave. After graduating from Harvard in 1914 at the age of 16, he obtained a graduate fellowship at Rice University in Houston, Texas, so he could work toward a Ph.D. Some teaching was involved, at which he proved inept, and he was treated as an outcast.

After various attempts to make a career for himself, Billy settled down to a kind of double life, earning a living in occupations that made no use of his exceptional gifts while occasionally impressing people with feats of rote memory or lightning calculation. He is credited with inventing the notion of the black hole, the result of a star collapsing into a dense object from which no light can escape, but there is no sign that he understood the physics needed to substantiate the idea. Though occasionally he still made the newspaper headlines, it was due to his weird social behavior. He was described as inadequate in the *New York Times*; he sued the newspaper unsuccessfully. Unemployed and destitute, Sidis died following a stroke in his forty-sixth year.

Leo, the father of Norbert Wiener, was an immigrant to the United States from Russia. He married a native-born American, Bertha Kahn, and they settled in Boston, where they soon got to know the Sidis family. Leo was a remarkable scholar, full of intellectual enthusiasm and energy. By the time their son Norbert was born, on 26 November 1894, Leo had been recruited to the Harvard faculty, where he rose to the rank of full professor in the department of Slavonic languages and literature. The Wieners had one other son and two daughters.

Influenced by the educational theories of his friend Boris Sidis, Leo regarded his son as an essentially average boy whose unusual abilities were the result of the intensive early education his father provided for him. The hot-tempered Leo was a hard taskmaster, more so than Boris; he was verbally abusive to his son and criticized him harshly when he failed to meet his high expectations. After failing to find a suitable school for Norbert, his father mainly educated the boy at home. Later, Norbert wrote:

> I began to discover that I was clumsier than the run of children around me. Some of this clumsiness was genuinely poor muscular coordination, but more of it was based on my defective eyesight . . . At school they viewed me socially as an undeveloped child, not as an underage adolescent. A further source of my awkwardness was psychic rather than physical. I was socially not yet adjusted to my environment and I was often inconsiderate, largely through an insufficient awareness of the exact consequences of my action. A further psychic hurdle I was to overcome was impatience. This impatience was largely the result of a combination of mental quickness and physical slowness. I had no proper idea of cleanliness and personal neatness, and I never myself knew when I was to blurt out some unpardonable rudeness or double entendre. I was already too much of a lone wolf.
>
> As the infant prodigy comes to realize that the elders of the community are suspicious of him, he begins to fear the reflections of this suspicion in the attitude of his contemporaries . . . Every child, in gaining emotional security, believes in the values of the world around him . . . He wishes to believe that his elders . . . are all wise and good. When he discovers that they are not, he faces

the necessity of loneliness and of forming his own judgment of a world that he can no longer fully trust. The prodigy shares this experience with every child, but added to it is the suffering which grows from belonging half to the adult world and half to the world of the children around him. Hence, he goes through a stage when his mass of conflict is greater than that of most other children, and he is rarely a pretty picture. (Wiener 1953)

The early development of Norbert's abilities was uneven, with his social competence lagging behind his intellectual skills. He was saddled with the job of bringing up his younger brother, and throughout his life nursed a grievance about this. The Wiener household had a fine collection of books, of which the boy took full advantage, despite poor eyesight. He was particularly attracted to zoology, physics, and chemistry, but in mathematics he was a late bloomer.

However, unlike the unfortunate Billy Sidis, Norbert was encouraged to have plenty of interests apart from his studies. For example, his father, whom he greatly admired, used to distract him from his studies by taking him for long walks. Even so, Norbert found life very difficult as a young man, largely because his overprotective parents had made it hard for him to become fully independent and strike out on his own. By late adolescence he should have left his parents' close-knit Jewish household, but his mother made it clear that she would never forgive him if he did. In his own words, "Like many other adolescents, I walked in a dark tunnel of which I could not see the issue, nor did I know whether there was any" (Wiener 1953).

Wiener was as successful as Sidis was not. Wiener thought the problem with Sidis was the way his proud father exposed him to the media: "The collapse of Sidis was in large measure his father's making." He believed that emotional starvation and parental exploitation warped Sidis's upbringing and that he reached manhood with a feeling of helplessness and inability to handle the practicalities of life, for which he blamed his parents. His education, Wiener said, produced a boy who could do wonderful, even brilliant reasoning in mathematics but had difficulty transferring that reasoning power to everyday affairs. Even among creative individuals who have good reasons to be grateful for their parents' efforts to help develop their intellects, a number have suffered from the shadows cast by the more negative effects that often arise when the parent is intent on making the child outstandingly able. Sidis's mother wrote a book (never published) called *How to Make Your Child a Genius.*

There is ample evidence that both Sidis and Wiener were affected by some disorder of the autistic spectrum, as their fathers probably had been. From what we know, we can infer that Sidis was the more severely affected, so much so that he never had much chance of making a success of life. Some other mathematical child prodigies, notably Ramanujan, also seem to have been affected by an autistic disorder. Child prodigies in mathematics occur quite frequently. We usually hear of them first when they enter university at a tender age; in many cases we do not hear of them again. Yet the mistaken idea that they just "burn out" seems to derive from rare cases such as that of the unfortunate Sidis. Let us look at a more recent example, whose life story has been described by Marjorie Senechal (2004).

Robert (Bobby) Ammann grew up near Boston, where his strict and impatient father had a job as an engineer. Bobby could read, write, and subtract by the time he was 3, but suddenly before he was 4, he stopped talking, for some unknown reason. For months only his mother could understand his mumbling. Gradually, with the help of a therapist, he started speaking again, but slowly. He moved slowly too. He was myopic and absent-minded; he never smiled and wouldn't play with other children. He was happiest on his own with copies of *Scientific American,* books on popular science, and dozens of mathematics textbooks. Schoolwork bored him, but his examination results were excellent. Both Harvard and MIT invited him to apply for admission but turned him down after interviews. Brandeis accepted him as a student but after three years asked him to leave. He then took a course in computer programming and got a job as a programming engineer at Honeywell. The work was so easy for him he could do it in his sleep. When he was laid off he kept on working until forbidden to enter the Honeywell premises. He then obtained a job in a post office, sorting mail. He lived in a motel convenient to his work and took meals at a nearby fast-food establishment.

During this period he became known to the mathematical community mainly through his discovery of aperiodic tilings independently of the better-known Roger Penrose. People remember him as kind and gentle but hard to relate to. He was neatly dressed, short and a little stout, his very high forehead framed by black hair and black-rimmed glasses. He didn't make small talk, or even say hello, and avoided eye-contact. He had a peculiar sense of humor and strange ideas about food. After Brandeis he didn't see much of his parents. As his mother said, their son "was off the charts intellectually, but impossible emotionally" (Senechal 2004).

Child prodigies are even more common in music than in mathematics. Mozart is the outstanding example, but there have been many others; recent examples include the Canadian pianist Glenn Gould, the Italian pianist Michelangeli, and the German conductor Carlos Schreiber. Again the right kind of upbringing is of vital importance, but achieving acceptance by the community seems less of a problem for musical prodigies. When musicians show autistic traits, the expressive power of music may help them suppress some of the symptoms. Of Béla Bartók, for example, it was said that when he played it was as if all music lived in him, but when he ceased playing he returned to the remotest depths of some cavern, from which he could be drawn only by force. There may be a parallel here with mathematics, in which people who exhibit autistic traits, especially impairment of social interaction, seem to have no difficulty composing and delivering excellent lectures—although they might be unusually direct with the audience, going straight in without the normal preliminary chitchat. Instead of relating to the audience, the speaker might talk almost as if to outer space and might skip many steps in mathematical proofs.

Age and Achievement in Mathematics

Hardy, in his testament *A Mathematician's Apology* (1967), states that "mathematics is a young man's game," and this aphorism has been much quoted. He remarks that Gauss is the only mathematician he can think of who published a major theorem in old age, and even then it was based on ideas Gauss had when he was much younger. Einstein (1951) said, "A person who has not made his great contribution to science before the age of forty will never do so" (James 2003b). Perhaps this is one of the reasons why the prestigious Fields medals are not awarded to mathematicians over the age of forty. Wiener (1953) largely agreed, writing that mathematics "is the athleticism of the intellect, making demands that can be satisfied to the full only when there is youth and strength." André Weil (1992) wrote, "Mathematical talent usually shows itself at an early age. There are examples to show that in mathematics an old person can do useful work, even inspired work, but they are rare, and each case fills us with wonder and admiration." To quote Hardy (1940) again, "If then I find myself writing, not mathematics but about mathematics, it is a confession of weakness, for which I may rightly be scorned or pitied by younger and more vigorous mathematicians. I write about mathematics because like any other mathematician who has passed sixty, I have no longer the freshness of mind, the energy, or the patience to carry on effectively with my proper job."

For scientists, including mathematicians, there is some literature on the subject of age and productivity. Of course productivity is not the same as creativity, which is hard to measure. The classic study by Lehman, reported in his *Age and Achievement* (1953), concluded that for scientists productivity climbs rapidly to a peak relatively early in life and then declines steadily and significantly. For mathematicians, he found that the peak occurred between 34 and 40. Nancy Stern (1978) considered the numbers of research articles published by mathematicians at various ages and evaluated them by number of indexed citations. She concluded that "the claim that younger mathematicians are more apt to create important work is unsubstantiated . . . I have found no clear relationship between age and achievement in mathematics." Some other investigators have found evidence that the productivity of scientists does not vary with age, but the most recent research supports the original thesis that there is a gradual decline. For example, Arthur Diamond (1986) has studied the productivity of research scientists in the United States between 1965 and 1979, using the number of indexed citations as a crude measure of quality. The survey covered full professors in the departments of mathematics at the Berkeley campus of the University of California and the Urbana-Champaign campus of the University of Illinois. Diamond concluded that quantity and quality of research output decline monotonically with age.

When Reuben Hersh sent out a questionnaire in 2001 to 250 mathematicians he knew (most were Americans) he received sixty-six replies, of which the more interesting are summarized in Hersh (2001). He describes Hardy's statement as misleading, even harmful, insofar as it may discourage people from mathematics when they are no longer young; it's unjustified and destructive. Hadamard and Erdös have proven that mathematicians can be productive into the eighth decades of their lives. The truth of the matter may be that mathematicians usually get their best ideas when they are young and go on to develop them later in life.

Chapter 3 } The Dynamics of Mathematical Creation

Mathematical Cognition

Cognitive activities are the active processes through which knowledge is acquired, such as perception, attention, memory, and learning. Interest in these activities has a long history, stretching back to classical times; today there is a vast literature on the subject. We cannot possibly cover this here, but fortunately a handbook of mathematical cognition has recently been published, containing twenty-seven essays on different aspects of the subject (Campbell 2005). The authors of the essays seek to throw light on such questions as:

- How does the mind represent numbers and make mathematical calculations?
- What underlies the cognitive development of numerical and mathematical abilities?
- What factors affect the learning of numerical concepts and procedures?
- What are the biological bases of number knowledge?
- Do humans and other animals share similar representations and processes?
- What underlies numerical and mathematical disorders, and what is the prognosis for rehabilitation?

Only some of the essays are directly relevant to the present work, but the reader who wishes for an up-to-date overview of research on mathematical cognition will find that the handbook provides it. Incidentally, the term "mathematics," in cognitive science, does not usually extend beyond simple arithmetic; much of the research is limited to young children.

Most abstract concepts are metaphysical in nature, drawing on the inferential structure of everyday bodily experience to reason about abstractions. Time, for example, is primarily conceptualized in terms of motion—either the motion of future times towards the observer or the motion of an observer toward a "time landscape." For centuries the mathematical concept of continuity was based on the idea of motion—the motion of a physical object with definite direction and speed. Such motion proceeds without gaps, interruptions, or discontinuities. For example, Euler described a continuous curve as "a curve described by freely leading the hand." This simple idea proved to be extremely rich and powerful and helped generate one of the most beautiful and productive branches of all mathematics: seventeenth-century calculus. Formally the mathematical function does not move, but cognitively the function does move, does approach limits. Further discussion of such matters may be found in *Where Does Mathematics Come From?* by Lakoff and Nunez (2000).

Cognitive styles are self-consistent modes of functioning, shown by individuals in their perceptual and intellectual activities, that correspond to habitual ways in which individuals organize and process information, solve problems, and make decisions. Although there are many variants, the two main cognitive styles are the intuitive and the logical, or analytic, styles. An intuitive person seeks to obtain a broad perspective on a problem and get an overall feel for it, reaching a conclusion fairly rapidly. An analytic person tends to take more of a logical step-by-step approach before deciding on a solution after a period of reflection. Neither cognitive style is generally preferable to the other, although one may be better than the other for certain tasks. It has been suggested that the left side of the brain is the analytic side, the right side of the brain the intuitive side, but this is an oversimplification. Most people use a mixture of cognitive styles, varying according to the matter under consideration.

For over a century psychologists have been interested in cognitive styles: they distinguish between visual thinkers and verbal thinkers, between intuitive thinkers and logical thinkers. It seems to be generally assumed that most creative mathematicians think mainly in pictures, although no survey to establish this conclusively has yet been carried out. We conjecture that the answer might be different for mathematicians working in different branches of the discipline, thus one might expect geometers to be visual thinkers, analysts to be verbal thinkers, and so on.

Even today, despite all that has been written on the subject, there does not seem to be a clear answer as to what proportion of mathematicians are mainly visual thinkers and what proportion are mainly verbal. Most mathematicians are intuitive thinkers, relying on the unconscious mind to a large extent, and there is some evidence to support this idea. In the case of mathematicians of genius, such as Poincaré, a good deal of research activity appears to take place at a relatively unconscious level. Although the difference between intuitive thought and logical Aristotelian thought was already recognized in the days of Plato, there is a need for further research.

In his *Enquiry into Human Understanding* (1748), the British empiricist David Hume wrote that to "join incongruous shapes and appearances costs the imagination no more trouble than to conceive the most natural and familiar objects" and that "this creative power of the mind amounts to no more than the faculty of compounding, transposing, augmenting, or diminishing the material afforded us by the senses and experience." Toward the end of the nineteenth century, von Helmholtz (1873) observed that "memory images of purely sensory impressions . . . may be used as elements of thought combinations without it being necessary, or even possible, to describe these in words," and further that "equipped with the awareness of the physical form of an object, we can clearly imagine all of the perspective images which we might expect upon viewing from this or that side, and we are immediately disturbed when such an image does not correspond to our expectations" (Helmholtz 1873). Galton, in his *Inquiries into Human Faculty and Its Development* (1883), was one of the first to realize that some people are verbal thinkers while others are visual thinkers. When he set out to investigate this, "To my great astonishment I found that the great majority of the men of science to whom I first applied protested that mental imagery was unknown to them . . . on the other hand when I spoke to persons I met in general society I found an entirely different disposition to prevail."

Galton believed that the matter merited proper research, but until recently relatively little has been done. Ann Roe (1951) made a study of imagery in sixty-one eminent American research scientists in the fields of biology, physics, psychology, and anthropology. She found that twenty-two were essentially visualizers, nineteen were essentially verbalizers, thirteen saw no images, and seven were mixed. The biologists were concentrated in the visual imagery group with the experimental physicists, anthropologists, and psychologists, while the theoretical physicists tended more to employ verbal or other symbolizations. Mathematicians were not included in the study. Interest in mental images has recently resurfaced with the work of Michel Denis (2004) in France, among others. Nowadays, in addition to visual imagery, three types are distinguished: verbal, auditory, and kinaesthetic. Galton wondered whether the type of image used might be hereditary; Roe obtained some evidence that this was so.

Many creative people claim to "see" the solution to a problem in an image. The physicists Michael Faraday and James Clerk Maxwell visualized electromagnetic fields as tiny tubes filled with fluid. Albert Einstein, who was very much a visual thinker, imagined what it would be like to ride on a beam of light or to drop a coin in a plummeting elevator. People with autism have problems learning things that cannot be thought about in pictures; even those who have good speech and the ability to articulate their thought processes mostly think in visual images. Temple Grandin (1996), who has autism, provides a vivid description in *Thinking in Pictures* (1996). As she says, one of the most profound mysteries of autism has been the remarkable ability of most autistic people to excel at visuospatial skills while performing so poorly at verbal skills. Galton himself was a visual thinker. He wrote:

> It is a serious drawback to me in writing, and still more in explaining myself, that I do not so easily think in words as otherwise. It often happens that after being hard at work, and having arrived at results that are perfectly clear and satisfactory to myself, when I try to express them in language I feel that I must begin by putting myself upon quite another intellectual plane. I have to translate my thoughts into a language which does not run very evenly with them. I therefore waste a great deal of time in seeking for appropriate words and phrases and am conscious, when required to speak on a sudden, of being very obscure through verbal maladroitness, and not through want of clearness of perception. (Galton 1833/1951)

Galois was said to have had difficulty in expressing himself verbally because he worked almost exclusively in his head, while Poincaré was a rather extreme example of an auditory thinker. Hardy used to say that for him thought was impossible without words. Alain Connes is quoted in Changeux and Connes (1995): "Expert mathematicians are endowed with a clairvoyance, a flair, a special instinct comparable to a musician's fine-tuned ear or a wine-taster's experienced palate that enables them to perceive mathematical objects directly. The evolution of our perception of mathematical reality causes a new sense to develop, which gives us access to a reality that is neither visual, nor auditory, but something else altogether."

We have already mentioned the inquiry into the working methods of mathematicians organized by Claparède and Flournoy. Responses to their questionnaire were received from almost one hundred mathematicians and others in Europe and the United States, most of whose names are unknown today. Not all respondents answered all questions, of which there were thirty-one, some in several parts. The results were published in a series of fifteen articles in the journal *L'enseignement mathématique* ("mathematical education"). One of the aims of the survey was to discover how mathematicians worked, for example, whether they preferred to work standing, seated, or lying down: several respondents, including Hadamard, said they preferred to work pacing up and down. Another question was whether the respondents believed an aptitude for the kind of abstract thought required for mathematics was in any sense inherited, and if so how. This produced seventy-six replies; over a third were negative, while the remainder were too vague or too brief for any firm conclusions to be drawn.

Only a few of the questions in the Claparède-Flournoy inquiry were of much psychological interest. About five respondents replied affirmatively to one that asked whether they experienced mood swings. Another question read: "It would be very helpful for the purpose of psychological investigation to know what internal or mental images, what kind of 'internal world,' mathematicians make use of: whether they are motor, auditory, visual or mixed, depending on the branch of the discipline which they are studying." Only twenty-six replies were received, and four were negative. Of the remainder, twelve said they were visual thinkers, two auditory, one graphic, one verbal-motor, and six mixed. On the whole, the results of this early survey were rather unilluminating. The respondents did not include David Hilbert, whose working methods were observed by Courant (1981):

> He spent his whole time gardening and in between gardening and little chores, he went to a long blackboard, maybe twenty feet long, covered so that also in the rain he could walk up and down, doing his mathematics in between digging some flower beds. All day one could observe him. I happened to have a student room on the fifth floor from which I could look out of the window and see Hilbert in his garden. He had a bicycle and practised little stunts on it.

It seems difficult to obtain useful information through questionnaires, but in the past ten years there have been some successful exercises.

Poincaré's disciple Hadamard was very interested in such matters as mathematical cognition and lectured about them in Paris in 1937 and later in New York. Then, in 1944, he wrote up these lectures for publication in his fascinating memoir, *The Psychology of Invention in the Mathematical Field*. This includes a translation into English of the Claparède-Flournoy questionnaire and of parts of Poincaré's famous lecture of 1908 on mathematical invention (for an English translation of the complete lecture see Halsted [1946]). Hadamard corresponded about the questions raised by Poincaré with eminent mathematicians including George Birkhoff, Norbert Wiener, Jessie Douglas, George Pólya, and Albert Einstein. Hadamard believed that his conclusions applied just as much in other fields of the arts and sciences.

Hadamard himself insisted that when he was engaged in thought, words were absent from his consciousness, although "it is quite possible, and rather probable, that words are present in fringe consciousness, such is the case I imagine for me, as concerns words used in mathematics . . . I insist that words are totally absent from my mind when I really think . . . even after reading or hearing a question, every word disappears the very moment that I am beginning to think it over." Not all his correspondents gave such answers, however. Often the mathematicians reported that their thought processes involved mental words, algebraic signs, and various vague images. Birkhoff reported that he visualized algebraic signs; Wiener that he thought both in words and without them; Douglas emphasized not the significance of words as such but the rhythm of their enunciation, with an emphasis on separate syllables, similar to the tapping out of the Morse code. Pólya characterized his own thinking as verbal. "I believe," he wrote, "that the decisive idea which brings the solution of a problem, is rather often connected with a well-turned phrase or sentence. The phrase or sentence enlightens the situation, gives things, as you say, physiognomy" (Hadamard 1945).

Einstein's responses were quite definite. To the question, " 'What internal or mental images, what kind of internal images—word, motor, auditory, visual or mixed—do mathematicians use?' Einstein answered that in his case the images were of the visual or, sometimes, motor type. In reply to the same question, but this time concerning not mathematical but ordinary thought, Einstein answered that in that case too his images were visual and motor; as for words, when they intervened at all, they were 'purely auditive.' He added that he was not satisfied with the answers he had given and that he would like to be asked more questions 'if it would be of any advantage for the very interesting and difficult work undertaken by Hadamard' " (Hadamard 1945).

Einstein's statements about his thought mechanisms attest to the great importance of images in the process of creative thought. "Words or language," he said, "whether written or spoken, do not seem to play any part in my thought processes. The psychological entities that serve as building blocks for my thought are certain signs or images, more or less clear, that I can reproduce or recombine at will." During the first stage of problem-solving, the essential feature for Einstein was the "combinatory play of images" directed by the desire to arrive at logically connected concepts. The fact that words intervene in a secondary stage has been confirmed by many investigators, yet whether words are completely absent in the first stage remains unproven.

In his *Autobiographical Notes* (1951), Einstein comments further about the formation of wordless concepts arising out of our normal perceptions of causal connections and relationships. The effect of these concepts, he says, is particularly evident when we encounter something out of line with our previous ideas, causing feelings of wonder or marvel. As an example of such a marvel, Einstein describes being shown a compass by his father at the age of 4 or 5 and being amazed by the compass needle, which functioned without being touched. Einstein believed that in his unconscious world of concepts action was associated with touch, and since touch was absent in the case of the compass, it produced the feeling of wonder that he regarded as an important stimulus for human intellectual development.

The Unconscious Mind

One of the basic findings of cognitive science is that most thought is unconscious. Ideas are not purely abstract, disembodied, transcendental entities but arise from and are shaped by the structure of human bodies and brains and the nature of everyday human experience. In mathematical invention, intuition plays an essential role. An idea arrives apparently out of the blue, except that it is relevant to the particular problem the person has been thinking about. There is no argument; it just feels right. Poincaré distinguished between two mental mechanisms: one logical, capable of the intensive work on a problem that usually precedes an intuitive breakthrough and the calculations that follow it, which he identified with the conscious mind; the other closely linked to the aesthetic sense and capable of recognizing the pattern, among all the possibilities that present themselves, that is both beautiful and important. The resolution of the pattern that solves the problem, he says, cannot be willed but comes of its own accord in what seems like a sudden flash of intuition from the unconscious mind.

Poincaré was very ready to discuss his own experiences with psychologists and to compare them with the experiences of others, as he did in his famous lecture when he recalled how he made the discoveries that made him famous. He had been thinking about periodicity with respect to linear fractional transformations of a certain class of functions (the reader may disregard the technical terms in what follows):

> For fifteen days I struggled to prove that no functions analogous to those I have since called Fuchsian functions could exist; I was then very ignorant. Every day I sat down at my work-table where I spent an hour or two; I tried a great number of combinations but arrived at no result. One evening, contrary to my custom, I took black coffee; I could not go to sleep; ideas swarmed up in clouds; I sensed them clashing until, to put it so, a pair would hook together to form a stable combination. By morning I had established the existence of a class of Fuchsian functions, those derived from the hypergeometric series. I had only to write up the results, which took me a few hours. (Hadamard 1945)

This seeming awareness of the workings of one's own unconscious mind is called metacognition by psychologists, and it appears to be a fairly rare phenomenon. Poincaré goes on to describe what happened one day shortly afterward when he went on a field trip organized by the École des Mines:

> The exigencies of travel made me forget my mathematical labours; reaching Coutances we took a bus for some excursion or other. The instant I put my foot on the step the idea came to me, apparently with nothing in my previous thoughts having prepared me for it, that the transformations I had used to define Fuchsian functions were identical to those of non-Euclidean geometry. I did not make the verification; I should not have had the time, because once in the bus I resumed an interrupted conversation; but I felt an instant and complete certainty. On returning to Caen, I verified the result at my leisure to satisfy my conscience. I then undertook the study of certain arithmetical questions without much apparent success and without suspecting that such matters could have the slightest connexion with my previous studies. Disgusted at my lack of success, I went to spend a few days at the seaside and thought of something else. One day, while walking along the cliffs, the idea came to me, again with the same characteristics of brevity, suddenness, and immediate certainty, that the

transformations of indefinite ternary forms were identical with those of non-Euclidean geometry.

On returning to Caen, I reflected on this result and deduced its consequences; the example of quadratic forms showed me that there were Fuchsian groups other than those corresponding to hypergeometric series; I saw that I could apply to them the theory of thetafuchsian functions, and hence that there existed thetafuchsian functions other than those derived from the hypergeometric series, the only ones I had known up to then. Naturally I set myself the task of constructing all these functions. I conducted a systematic siege and, one after another, carried out all the outworks; there was however one which still held out and whose fall would bring about that of the whole position. But all my efforts served only to make me better acquainted with the difficulty, which in itself was something.

At this point I left for Mont-Valérien, where I was to discharge my military service. I had therefore very different preoccupations. One day, while crossing the boulevard, the solution of the difficulty which had stopped me appeared to me all of a sudden. I did not seek to go into it immediately, and it was only after my service that I resumed the question. I had all the elements, and had only to assemble and order them. So I wrote out my definitive memoir at one stroke and with no difficulty.

Poincaré goes on to reflect on these experiences:

Most striking at first is this appearance of sudden illumination, a manifest sign of long unconscious prior work. The role of this unconscious work in mathematical invention appears to me incontestable, and traces of it would be found in other cases where it is less evident. Often when one works at a hard question, nothing good is accomplished at the first attack. Then one takes a rest, longer or shorter, and sits down anew to the work. During the first half-hour, as before, nothing is found, and then all of a sudden the decisive idea presents itself to the mind. It might be said that the conscious work has been more fruitful because it has been interrupted and the rest has given back to the mind its force and freshness. But it is more probable that this rest has been filled out with unconscious work and that the result of this work has afterwards revealed itself to the geometer just as in the cases I have cited.

There is another remark to be made about the conditions of this unconscious work: it is possible, and of a certainty it is only fruitful, if it is on the one hand preceded by and on the other hand followed by a period of conscious work. These sudden inspirations never happen except after some days of voluntary effort which has appeared absolutely fruitless and whence nothing good seems to have come, where the way taken seems totally astray. These efforts then have not been as sterile as one thinks; they have set going the unconscious machine and without them it would not have moved and would have produced nothing. (Hadamard 1945)

Unlike metacognition, the type of experience Poincaré describes here seems quite common. Poincaré's complete confidence in the power of his unconscious mind may be unparalleled, but other mathematicians have recorded instances of the role of the unconscious in their creative work. The Norwegian mathematician Niels Abel once solved a problem in his head while lying in bed, but he then forgot the sequence of proofs upon which he had built his conclusion. No matter how much he brooded over this, he was unable to recover his steps. Finally one night, Abel woke from sleep with a whoop of joy. The solution and the whole sequence had, in a flash, become evident to him. This was simply a case of something being recollected during sleep, and it is much more remarkable when the solution to a problem that has defeated the best efforts of the thinker appears quite suddenly and spontaneously after a period of sleep. Gauss, referring to an arithmetical theorem that he had tried unsuccessfully to prove for years, gave this example:

> Finally, two days ago, I succeeded not on account of my painful efforts, but by the grace of God. Like a sudden flash of lightning, the riddle happened to be solved. I myself cannot say what was the conducting thread which connected what I previously knew with what made my success possible. It is unnecessary to observe that what happened to me on my awakening is perfectly similar and typical, as the solution which appeared to me was without any relation to my attempts of former days, so that it could not have been elaborated by my previous conscious work and appeared without any time for thought, however brief. (Hadamard 1945)

Psychologists have recognized this suddenness and spontaneity as a common hallmark of many kinds of invention. Many scientists have offered anecdotes about their own experience with this phenomenon; Helmholtz commented on it, as did the chemist Wilhelm Ostwald, the technologist Nikola Tesla, and the physicist Paul Langevin. When asked about his working methods, the Russian mathematician Andrei Nikolaevitch Kolmogorov answered:

You read some books, to prepare your own lectures with some new variations or other, and suddenly, from the soil of this everyday work, some unexpected idea emerges and vaguely as yet, some completely different approach can be seen. Then having worked it out, almost everything else is neglected—and one thinks endlessly along the lines which have just appeared. Fortunately I usually had the opportunity to do this, but in the whole history of my scientific discoveries such complete oblivion, cut off from everything else, might last for one week, sometimes possibly for two—not more. (McFaden 2000)

Kolmogorov said he preserved a lively interest in a problem only as long as it was still unclear whether the problem would be solved or not. As soon as this became clear, he would try to hand the whole thing over to an apprentice to write up the proofs. Due to old age and laziness, he said, having done something good, "At best I write it immediately but usually throw away attempts at amplification and extension" (McFaden 2000). Kolmogorov thought that the way mathematical ideas came to him was similar to the way musical ideas came to Mozart.

The role of intuition in mathematical research is not fully appreciated outside of the profession. In his memoir, Hadamard discusses what he calls "paradoxical cases of intuition," with illustrations from the experiences of Fermat, Galois, and Riemann. The case of the Indian mathematician Ramanujan provides an extreme example of intuitive thought. "His ideas of what constitutes a mathematical proof were of the most shadowy description," explained his mentor Hardy. "All his results, new or old, right or wrong, had been arrived at by a process of mingled argument, intuition and induction, of which he was entirely unable to give a coherent account" (Hardy 1940). Ramanujan said that while he was asleep the Hindu goddess Namagiri would appear to him in a dream and tell him wonderful formulae (Kanigel 1991).

Hadamard stated that the earlier findings of Helmholtz and others were in accordance with his own experience. He distinguished four phases in the process of mathematical creation. The first is "preparation," which involves the trial and error the mathematician piously omits when he presents his work in polished form. Two further stages, called "incubation" and "illumination," build upon the first. In the incubation stage the study seems to have been completely interrupted and is dropped. However, after the preparatory and incubation stages, the mathematician sometimes experiences a surge of mental images accompanied by a sudden sense of illumination that spreads through the brain and his entire being. This is the critical step in the work of mathematical creation, but a fourth and final stage called "verification," more conscious than the preceding two, necessarily follows: the deliberate process of definition that permits an argument, a theorem, or a proof to be stated precisely and then put to use. In this last step, reason and judgment are called into play (Hadamard 1945).

In reviewing Hadamard's memoir for the *Mathematical Gazette*, Hardy (1940) comments that the mystery lies entirely in the early stages of this process, and first in the initial stage of "preparation." It is plain, he writes, that during this stage, however futile it may have seemed, the mathematician has done something essential; he has shaken up his ideas in a way that somehow makes later illumination possible. To discover is to combine ideas fruitfully, and in the preparation stage the mathematician has formed a mass of combinations. These may have seemed useless or uninteresting; but the process of forming them is more productive than it may seem, since it sets in motion the unconscious machinery. However, the most puzzling question, Hardy concludes, is how we unconsciously select from among all these combinations.

Einstein is said to have suggested that the creative scientists are those with "access to their dreams." Wiener (1953), in his fascinating account of his childhood and youth, tells us this about his working methods:

> Granted an urge to create, one creates with what one has. With me, the particular assets that I have found useful are a memory of a rather wide scope and great permanence and a free-flowing, kaleidoscope-like train of imagination which more or less by itself gives me a consecutive view of the possibilities of a fairly complicated intellectual situation. The great strain on the memory in mathematical work is for me not so much the retention of a vast mass of fact in the literature as of the simultaneous aspects of the particular problem on which I have been working and of the conversion of my fleeting impressions into something permanent enough to have a place in memory. For I have found that if I have been able to cram all my past ideas of what the problem really involves into a single comprehensive impression, the problem is more than half solved. What remains to be done is very often the casting aside of those aspects of the group of ideas that are not germane to the solution of the problem. This rejection of the irrelevant and purification of the relevant I can do best at moments when I have a minimum of outside impressions. Very often these moments

seem to arise on waking up; but probably this really means that sometime during the night I have undergone the process of deconfusion which is necessary to establish my ideas. I am quite certain that at least a part of this process can take place during what one would ordinarily describe as sleep, and in the form of a dream. It is probably more usual for it to take place in the so-called hypnoidal state in which one is awaiting sleep, and it is closely associated with those hypnagogic images which have some of the sensory solidity of hallucinations, but which, unlike hallucinations, may be manipulated more or less at the will of the subject.

Changeux and Connes (1995) discuss Hadamard's memoir in their *Conversations on Mind, Matter, and Mathematics*. Connes concurs with the kind of sensations Hadamard experienced and adds that the process of verification can be very painful: "One is terribly afraid of being wrong. Of the four phases it involves the most anxiety, for one never knows if one's intuition is right—a bit as in dreams, where intuition very often proves mistaken. But the moment illumination occurs, it engages the emotions in such a way that it's impossible to remain passive or indifferent."

The Impact of Personality and Developmental Disorders on Mathematical Creativity

Anthony Storr's classic survey *The Dynamics of Creation* (1972), the title of which we have adapted for this chapter, provides a useful introduction to the vast literature on creativity. Much of the literature concerning creativity in other fields has some relevance to mathematics, just as much of what is known about invention in the mathematical field applies more generally. Storr begins by discussing the theories of Sigmund Freud and his followers on the causes of creativity. Psychoanalysts are among many researchers primarily concerned with these causes, whereas Storr and others are more concerned with what drives those who make full and effective use of their creative gifts. Storr believes that it is often a matter of personality more than anything else, and so devotes most of his book to a discussion of the relation between creativity and different types of personality, giving illustrations mainly from literature and music.

Personality and developmental disorders are deeply ingrained and enduring behavior patterns that manifest themselves as inflexible responses to a broad range of personal and social situations. They represent significant deviations from the way the average individual in a given culture perceives, thinks, feels, and, particularly, relates to others. Frequently, but not always, such disorders are associated with various degrees of subjective distress and problems of social functioning and performance. They appear in childhood and may persist throughout life.

Schizoid Personality Disorder

Storr begins with schizoid personality disorder. Wolff and Chess (1964) often found schizoid personality traits in the histories of schizophrenic patients, although they emphasized that this tells us nothing at all about the risk of schizophrenia in people with schizoid personality. Schizophrenia is a relatively rare condition, while schizoid traits may be quite common. Wolff (1995) explains that such traits, despite their association in a few people with serious psychiatric illness, "may be biologically advantageous in general because of their possible association with originality and giftedness." People with the disorder display the following clinical features: they neither desire nor enjoy close relationships, choose solitary activities, have little interest in sexual experiences, get pleasure from few activities, lack close friends other than near relatives, are indifferent to praise or criticism, are emotionally cold, and show detachment or flattened affectivity.

Detachment and emotional isolation characterize the schizoid personality. Often, but not invariably, an individual with this character structure gives an impression of coldness combined with an apparent air of superiority that is not endearing. He or she has problems taking an emotional interest in other people. Others get the feeling that such an individual is unconcerned with, if not superior to, the ordinary mundane preoccupations of average people and that he is out of touch with, or on a different wavelength from, the people with whom he or she mingles but does not mix. Very often the person with a schizoid personality is accused of keeping other people at arm's length and of avoiding intimacy—an accusation that is justified. Sometimes these individuals are said to be wearing masks—also an accurate observation, since an individual with a schizoid personality habitually plays roles that intellectually she believes to be appropriate but that do not reflect what she actually feels. Thus, she may decide that it is morally right to be generous or tactful or considerate and behave accordingly. Because, however, this behavior originated from an intellectual decision rather than from true feelings, it is likely that all that will be conveyed to the recipient of her attentions is an impression of exaggeratedly good manners. Such an individual lacks the personal touch—the feeling, if not of intimacy, at least of some shared common ground upon which one person meets another as a human being.

People with schizoid personality disorder tend to seek meaning and significance in things rather than in other people; this is highly relevant to scientific creativity. Because emotional involvement with others appears dangerous, they remain detached and isolated. A paradoxical characteristic of the disorder is that people with it have a sense of extreme weakness and vulnerability vis-à-vis others, combined with its exact opposite, a sense of superiority and potential omnipotence. People with schizoid personality disorder often fail to develop a realistic sense of their position in the human hierarchy, because at a very early stage they cease to interact genuinely with their peers. Thus they often feel characteristically weak and incompetent on the one hand, and have equally unrealistic fantasies of power on the other. The less satisfaction they gain by reacting to the external world, the more they become preoccupied with their inner world of fantasy. People with schizoid personalities are essentially introverted.

Storr suggests five reasons creative activity is an apt way for schizoid individuals to express themselves. First, since most creative activity is solitary, choosing such an occupation means that the schizoid person can avoid the problems of direct relationships with others: the social situation is under control. Second, creative activity enables the schizoid person to retain at least part of his fantasy of omnipotence by creating his own world. Third, creative activity reflects the characteristics of the schizoid personality because it places a greater importance on internal than external reality. Fourth, certain kinds of creativity allow the person with a schizoid personality to feel they can impose their own order on a world that hitherto seemed unpredictable. Finally, creative activity can act as a defense against the threat of finding the world meaningless and arbitrary.

Storr (1972) selected the physicists Isaac Newton and Albert Einstein as exemplars of the schizoid personality (also see James 2003a). Sula Wolff (1995) chose the philosopher Ludwig Wittgenstein, cautioning that his abnormal personality can in no way be regarded as the most important aspect of his productive life, nor as an explanation of his genius. Other authorities say that these and others are better regarded as exemplars of the Asperger personality, which has many features in common with the schizoid personality.

Asperger Syndrome

In psychiatry, a syndrome is a collection of symptoms with a common cause, such as a developmental or personality disorder. The symptoms of Asperger syndrome are generally grouped under six headings: impairments of social interaction, all-absorbing narrow interests, repetitive routines, speech and language peculiarities, problems of nonverbal communication, and often motor clumsiness. The disorder can manifest itself in a bewildering variety of ways; not all individuals will exhibit all of the symptoms, although for the diagnosis to be made at least one symptom from under most of these broad headings should be present. The disorder usually shows itself in early childhood. It is estimated that it affects about one in two hundred of the general population, with a much higher proportion of males than females. The syndrome is much more common than classical autism, in which the individual seems trapped in a private world.

Hans Asperger, one of the pioneers in the study of autism, was a Viennese pediatrician who described the disorder to which his name has been attached and observed: "To our own amazement, we have seen that autistic individuals, as long as they are intellectually intact, can almost always achieve professional success, usually in highly specialized academic professions, often in very high positions, with a preference for abstract content. We found a large number of people whose mathematical ability determines their professions" (Asperger 1944). It is now well established that people with Asperger syndrome are attracted by mathematics and kindred subjects such as computer science. Their greatest wish is to bring the world under the control of pure reason, to create order and meaning out of the chaos they experience around them, particularly in the puzzling social domain. However, they lack the capacity to experiment in the social world and to produce empirical information to validate or refute their models.

The syndrome is not universally recognized in psychiatry, but gradually seems to be gaining acceptance. Although it is a relatively mild form of autism, its effects are by no means mild. It is not uncommon for individuals to have only a few features of the syndrome rather than the full profile.

Those affected by it can often live fairly "normal" and successful lives, but it must be emphasized that among people with the syndrome, only a fortunate minority have both the ability and the opportunity to excel.

Typical "Asperger traits" include restricted, repetitive, stereotyped, and obsessive patterns of behavior, interest, and activities; impaired social interaction and lack of empathy; imposition of routines and control on self and others; naivety, childishness, and lack of common sense; unusual sense of humor and use of language; difficulties with nonverbal communication, including reading the facial expressions, vocal inflections, and body language of others; and clumsiness or difficulty with motor skills. For example, a person with Asperger syndrome might rock back and forth when talking or while seated. She might run rather than walk. While walking, she might move her arms very little and have a somewhat awkward gait. Other noticeable features of the syndrome are an avoidance of eye contact, reduced general movements of the limbs or body, and a lack of social contact with other people in the room. People with Asperger syndrome sometimes exhibit stereotyped movements of the body, such as flapping the arms slightly. There seems to be some connection between music and autism spectrum disorders. In therapy, it has been found that in some cases autistic traits disappear temporarily when music is played. Oliver Sacks (1995) gives an example: perhaps Einstein experienced this when he played the violin.

Although the cause of Asperger syndrome, and of other autistic disorders, is not fully understood, genetic factors are certainly involved. It is hereditable; there is usually some trace of the syndrome in the forebears of someone who has it. Since autism has been recognized by psychiatrists only within the past sixty years, numerous past cases must have gone unrecognized, although it may seem surprising that recent biographers should pass over what must be one of the main features of the life stories of their subjects. The syndrome is not properly understood by many otherwise well-informed people, who find it hard to realize what some people with this disorder may be capable of achieving. It should be emphasized that disorders on the autistic spectrum are disorders of development, whose effects are present throughout life. They are not mental illnesses, like schizophrenia or bipolar disorder, that have an onset in adolescence or later (though these are often combined with disorders on the autistic spectrum).

To assess Asperger syndrome, a psychiatrist considers the whole history of the patient from birth, conducts various standard tests, and uses his or her experience to arrive at a diagnosis. Obviously, it is impossible to carry out these procedures fully on historical figures who are no longer alive. In this situation, some writers on the subject refuse to accept the diagnosis, while others do so but with strong reservations. When we accept it, with due caution, for famous mathematicians in our profiles, we follow the precedent set by authors of standard books on autism who identify possible cases of Asperger syndrome, often from far back in history. Uta Frith gives some examples in her book *Autism: Explaining the Enigma* (2003). Michael Fitzgerald gives more case studies in his books *Autism and Creativity* (2004), *The Genesis of Artistic Creativity* (2005), and *Unstoppable Brilliance* (2006), and in various articles. In those publications the cases of Hamilton, Ramanujan, and Kurt Gödel are considered in some detail. Most of the subjects of this book's profiles exhibit Asperger traits; on the basis of the information available, Hamilton, Hardy, Ramanujan, Wiener, Dirac, and Gödel are good examples of people who satisfy the accepted criteria for Asperger syndrome. Temple Grandin, who has autism herself, identifies people she believes had Asperger syndrome in her *Thinking in Pictures* (1996). Some laypeople have a special interest in the field; for example, Norm Ledgin has suggested some Asperger possibles in his *Asperger's and Self-Esteem,* while Ioan James has combed the literature for others in his *Asperger's Syndrome and High Achievement* (2005) and in his article "Singular Scientists" (2003b).

Many features of Asperger syndrome enhance creativity. The ability of people with Asperger syndrome to focus narrowly on a topic and resist distraction is particularly important. When Isaac Newton was asked how he conceived the theory of gravitation, he replied "It was through concentration and sheer dedication. I keep the subject constantly before me, till the first dawning opens slowly, little by little and little into the full and clear light" (Keynes 1995). People with the syndrome show a remarkable capacity for persistence. They often reject received wisdom and the opinions of experts. They live very much in their intellects, and certain forms of creativity benefit greatly from this, particularly mathematical creativity.

As we have explained in the introduction, Asperger was struck by the attraction of mathematics for the people he was studying. The reader may well wish to be presented with stronger evidence of the connection between autism and the sciences, particularly mathematics. Fortunately, this has recently become available. Baron-Cohen (2001) has devised a questionnaire for measuring traits associated with the autistic spectrum in adults with normal intelligence. From the answers to the questionnaire a number is obtained, called the autistic-spectrum quotient, which gives an estimate of where a given individual falls on the continuum from "normality" to autism. When the questionnaire was given to 4,175 students at Cambridge University, 20.1 percent returned it, with no significant difference in this rate between disciplines. Scientists (including mathematicians) scored significantly higher on the questionnaire than both humanities and social sciences students, confirming the general belief that autistic traits are often associated with scientific skills. The mean scores (out of 50) were 18.5 in all sciences, as compared with 16.7 in humanities and 16.4 in social sciences. Within the sciences, the mean scores were 14.9 for biological sciences, 21.1 for computer science, 17.9 for engineering, 21.5 for mathematics, 15.4 for medicine, 19.6 for physical science, and 18.5 for nonspecific science. Scientists scored higher than nonscientists, and within the sciences, mathematicians, physical scientists, computer scientists, and engineers scored higher than the more human- or life-centered sciences of medicine and biology. Further investigation has revealed that a disproportionate number of mathematics students have received a diagnosis of autism, and that autistic individuals tend to have an unusually high proportion of engineers in their families. Baron-Cohen believes that mathematical talent is a sign of being able to systemize and that this systemizing ability, often developed to extremes, could be a hallmark of autism. Still more recent research by Baron-Cohen and his associates is reported in an article, "Mathematical Talent Is Linked to Autism," to appear in the journal *Human Nature*.

The Impact of Mood Disorders

Mood (or affective) disorders seem no more common among mathematicians than they are among the population in general. Two mathematicians we profile, Ada Byron and Georg Cantor, clearly had bipolar disorder. Some of our other subjects, notably Lagrange and Sonya Kovalevskaya, appear to have suffered from depression alone. Other mathematicians with depression, it has been suggested, were Richard Courant, Felix Klein, and Emil Post.

Only the mild forms of these disorders can be conducive to creativity. Typical depressive symptoms include apathy, lethargy, hopelessness, sleep disturbance (sleeping far too much or too little), slowed physical movements, slowed thinking, impaired memory and concentration, and a loss of pleasure in normally pleasurable events. In bipolar disorder, mood shifts dramatically between depression and mania, a state of elevated mood. The mild form of bipolar disorder is called cyclothymia; the mild form of mania is called hypomania. Although depression is not conducive to creativity, creative work can be a particularly effective way of protecting the self from the threat of an underlying depression. Even the creative person with bipolar disorder who cannot manage to stay within the milder states may find that some productive work can still be done in the more intense stages of depression and mania. However, mania can seduce the individual from creativity by increasing the desire to enjoy social life.

Hypomanic or manic individuals usually have an inflated sense of self-esteem as well as a certainty of conviction about the correctness and importance of their ideas. This grandiosity can contribute to poor judgment, which in turn often results in chaotic patterns of personal and professional relationships. During hypomania and mania, mood is generally elevated and expansive or else paranoid and irritable; activity and energy levels are greatly increased, the need for sleep is decreased; and speech is often rapid, excitable, and intrusive, moving quickly from topic to topic. In its milder variants, the increased energy, expansiveness, risk-taking, and fluency of thought associated with hypomania can result in highly productive periods.

The American mathematician Morris Kline has written an interesting account of how to make the best of bipolar disorder. He describes a symptom of mania, the flight of ideas, that can assist in creation: "Approaches and ideas are likely to occur with such rapidity and suddenness that one cannot pursue each one seriously at the moment. A good thing to do is to jot down these ideas so as not to lose sight of them." Such flights of ideas occur only during mania and are least chaotic during milder states. A depressed state, Kline goes on to explain, affects one's ability to think. One can force oneself to work only on something that is more routine or that really was worked out before and needs revision. Here Kline refers to something more intense than mild depression, for in that state creative thought may still be possible. Milder depression can be quite fertile and much less painful. It may also be a suitable state for polishing work and for carrying out a large-scale critical review that may point the way to new directions for growth. Some creative people with bipolar disorder have learned by trial and error to fit the phase of their work to the mood they are in. It helps if the individual has regular mood cycles and can discover what time of day, month, or year is most likely to bring depressions or manias. Some have several pieces of work-in-progress at different stages of completion, so that when depression arrives, in which work can be corrected, something is in progress that needs correction, and when mania arrives, bringing its cornucopia of ideas, a project in need of them awaits.

Mild versions of these mood disorders are quite common. Hypomania and cyclothymia may be productive states because they increase both the quantity of created work and its originality. These milder states can allow the individual to be more disciplined and less impulsive than more severe stages do, thus improving the chances of carrying work to completion. They induce less of the impatience and distractibility that interfere with work. Patience is one of the contributions hypomania makes to quality. The hypomanic works quickly, but attends to details as well as larger factors. The individual is inventive and resourceful but does not take as many foolish chances as someone in a manic state. He is also freer of the delusions of mania, better able to be realistic about his work and correct it.

Some individuals take on the less productive phases of their work as challenges to ingenuity and, regardless of mood, are always looking for ways to stimulate creativity. One means they employ is to try to increase the input on which their work depends. As Kline suggests, "Reading related material may be the best way to get the mind started on a new channel of thought and because the reading is related, this new thought may be the right one." The person who fluctuates between mild depression and mild mania profits from the best of both states. He or she is imaginative, original, insightful, conscientious, and willing to keep working until no further improvement can be made. The result is likely to be rich, deeply felt, of great range and scope, balanced between strength and subtlety. Mild versions of the mood disorders, such as mild depression or mania, are so common in the arts and sciences that they may pass unnoticed.

People with Asperger syndrome can also be moody, and their mood at times can fluctuate up and down. This might be confused with a mood disorder, but these conditions are radically different. First of all, autism always has an onset under three years of age, while bipolar disorder often has an onset in adolescence or later. People with bipolar disorder do not have the impairments in nonverbal behavior seen in Asperger syndrome, do not necessarily fail to develop peer relations, and do not lack spontaneous sharing of enjoyment and interests (quite the opposite: during the elated phase they are often overeager to be involved with other people). Lack of social and emotional reciprocity is not a feature of bipolar disorder; neither is encompassing preoccupation with at least one stereotyped and restricted pattern of interest, or apparently inflexible adherence to nonfunctional routines, or preoccupation with stereotyped movements or with parts of objects.

Severe depression is one of the causes of suicide, often at the recovery stage. Suicide seems to be quite rare among mathematicians, although no statistics are available. Only two recent cases come to mind, though there must have been more: suicide tends to be hushed up. One was the young Japanese mathematician Yukata Taniyama, who collaborated with Goro Shimura to devise the conjecture that played such an essential role in the proof of Fermat's Last Theorem. He left a suicide note that read: "Until yesterday I had no intention of killing myself. But more than a few must have noticed that lately I have been tired both physically and mentally. As to the cause of my suicide I don't quite understand it myself, but it is not the result of a particular incident, nor of a specific matter. Merely may I say I am in the frame of mind that I lost confidence in my future." Another recent suicide was that of the gifted young German mathematician Andreas Floer. The death of Alan Turing was probably suicide, and Kurt Gödel may be said to have ruined his health by semistarvation. We are aware of no more examples of completed suicide among prominent mathematicians, although both Hardy and Ramanujan made attempts to take their own lives.

Part II

TWENTY MATHEMATICAL PERSONALITIES

Chapter 4 } Lagrange, Gauss, Cauchy, and Dirichlet

Joseph-Louis Lagrange

On 25 January 1736, Joseph-Louis Lagrange was born in a small town near the city of Turin, at that time capital of the Piedmont and the seat of the Savoyard kings of Sardinia. He was one of eleven children, only two of whom reached maturity. Their father held the office of treasurer of constructions and fortifications in Turin but had lost most of his money through speculation. Their mother was the daughter of a physician and a member of the wealthy Conti family.

Lagrange's father intended his son to become a lawyer. Lagrange raised no particular objections to this plan, but eventually decided he would prefer to study the exact sciences instead. He did so with such success that by the age of 19 he had been appointed professor of mathematics at the Royal School of Artillery in Turin. In his youth, his temperament was called mild and melancholic, and it was said that he knew no other pleasure than study.

In 1763 Lagrange made his first visit to Paris; until then he had traveled little. He was received with honor but became seriously ill during his stay and returned home instead of continuing to London as he had planned. Thanks to Jean le Rond d'Alembert, then permanent secretary of the Paris Academy, Lagrange received an attractive offer from Frederick the Great of Prussia, expressing the wish of the "greatest king in Europe" to have the "greatest mathematician in Europe" resident at his court; in 1766 Lagrange became director of mathematical physics at the Berlin Academy. Before leaving Turin for the Prussian capital he made another visit to Paris but again fell ill after a banquet in his honor and departed without regret.

The next year Lagrange married a cousin, Vittoria Conti. In reply to an inquiry from d'Alembert, he wrote:

> I don't know whether I calculated ill or well, or rather, I don't believe I calculated at all; for I might have done as Leibniz did, who, compelled to reflect, could never make up his mind. I confess to you that I never had a taste for marriage . . . but circumstances decided me to engage one of my young kinswomen to take care of me and all my affairs. If I neglected to inform you it was because the whole thing seemed to me so inconsequential in itself that it was not worth the trouble of informing you of it.

However, the marital relationship deepened over the years, and when Vittoria died sixteen years later after a lingering illness, Lagrange was heartbroken.

During his twenty years at the Berlin Academy, Lagrange worked on what he called *mécanique analytique* (analytical mechanics), the application of calculus to the motion of rigid bodies. His conclusions were organized into a volume under that title published in 1788. This was his masterpiece— a scientific poem, according to Hamilton. He was exceedingly fastidious with regard to the mathematical form of his writings: "I have a bad habit," he wrote to d'Alembert, "which I am unable to shake off. I often rewrite my articles, often many times, until I am passably satisfied with them." He was a consummate artist, and his standard of mathematical elegance was exceptionally high.

Lagrange was already somewhat nervous in disposition and inclined to melancholy, but around 1780 he developed a major depression and lost interest in mathematics for some years. "I begin to feel the pull of my inertia increasing little by little, and I cannot say I shall still be doing mathematics ten years from now," he wrote to d'Alembert. "It also seems to me that the mine is already too deep, and that unless new veins are discovered it will have to be abandoned." D'Alembert wrote back: "In God's name do not renounce work, for you the strongest of all distractions. Good-bye, perhaps for the last time. Keep some memory of the man who of all in the world cherishes and honours you the most."

Following the death of Frederick the Great in 1786, an indifference towards science and a resentment of foreigners arose in Berlin. The following year Lagrange left the city, after twenty very successful years, to become *pensionnaire vétéran* of the Paris Academy, of which he had been a foreign

associate member since 1772. Because his father's family came from Tour-
aine (they were related to Descartes), he considered himself as much French
as Piedmontese. In the French capital he was received with every mark of
distinction; apartments in the Louvre were set aside for his reception and
Queen Marie-Antoinette doted on him. The Parisians found him gentle and
agreeable but unwilling to take a lead in conversation.

Soon he fell into depression again. He became absent-minded and melancholy. At social and scientific gatherings he would stand staring absently
out of the window, his back to the guests who had come to do him honor, a
picture of sad indifference. He told his friends and colleagues that mathematics was no longer important to him. If informed that some mathematician was engaged in important research he would say "so much the better; I
began it; I shall not have to finish it." It is said that the second edition of the
Mécanique analytique lay unopened on his desk for two years.

In 1792, at the age of 56, Lagrange took as his second wife the daughter of
a friend and colleague of his, and she gradually helped him regain his
interest in mathematics and in life in general. Like his first marriage, the
second was childless: Lagrange had no wish for children. During the revolutionary period he managed to remain politically neutral, although he was
granted a pension by the revolutionary government and served on the
commission charged with the establishment of standards for weights and
measures, out of which the metric system emerged. In 1795 he was appointed professor of mathematics at the new École Normale, and after that
temporarily closed down he became professor at the École Polytechnique.

Lagrange was a poor teacher; too indrawn, more adept at following
steadily and quietly his own ideas than in educating young men. Bonaparte
consulted him frequently and appointed him senator, then later count of
the Empire and grand officer of the Legion of Honor. The emperor described him as the "proud pyramid" of the mathematical sciences, but
George Sarton, writing about Lagrange's personality, remarked that he was
more like an iceberg than a pyramid.

Lagrange was of medium height and slight build, with pale blue eyes and
a colorless complexion. Among the fundamental traits of his character were
his peacefulness, modesty, and sweet temper. He was known for his gentle
demeanor and his diplomatic skills. It was said that he hardly ever hurt
anybody and instinctively avoided every quarrel or anything that might lead
to a quarrel. Modest and diffident to a fault, he detested controversy, and to

avoid it allowed others to take the credit for what he himself had done. At the same time he could be remarkably obstinate. His instinct for self-preservation often took the form of a kind of blindness or callousness, less a defense of self than of the task to be done. Lagrange was described as having a calm and pensive look, his glances penetrating but gentle. He spoke slowly, rolling a little his *r*'s, lifting his voice when he discussed a matter of interest to him. He did not care for rambling conversations and cared even less for idle ones.

Lagrange did not lack literary imagination or sensibility, but such thoughts were overwhelmed by the mathematical cogitations dominating his mind. "I love music because it isolates me," he wrote. "I listen to the first three bars, with the fourth bar I am already lost; I give myself up to my own reflections, nothing interrupts me and in that way I have solved many difficult problems." Lagrange was too self-centered to make many friends, although he was close to d'Alembert and maintained an extensive correspondence with other mathematicians. He managed to get rid of everything that might hamper his own studies. He took refuge in silence, and some people admired him the more for it. He was first and last a mathematician, grudging the time and energy given to any problem but the mathematical one that engrossed his mind.

Lagrange worked with great concentration. He was generally pensive and silent, but sometimes his thinking would throw him into a kind of feverish excitement, and his pulse would become irregular. He was very careful about diet and regimen; his interest in such matters caused him to read medical books, to study the nature of poisons and other dangerous substances, and to form physiological opinions of his own. His sedentary and studious life exposed him to hemorrhoids and to a kind of overheating. At the end of each winter he suffered from bilious fluxions and spells of hypochondria. "The spirit is lazy," he used to say. "One must always be on one's guard to prevent its natural slackness and to develop its natural strength for the future."

Lagrange acquired from Frederick the Great the habit of always doing the same things at the same times, finding that work thus regulated became easier and more agreeable. He led a ritualized life. In the morning he rose about ten o'clock and took coffee. He would then settle down to reading and writing letters. Since his stomach was easily upset, he ate little meat and never drank wine except in small amounts, diluted with water. Immediately after dinner he devoted a few hours to making visits or to his accustomed walk. He preferred to walk in the open air and to be alone when he did so, since walking helped him to meditate. At six in the evening he returned to his cabinet, where he shut himself in, that he might be sure of remaining undisturbed. At ten in the evening he drank tea, very hot and sweet, sometimes with lemon; he went to bed about midnight.

Lagrange's memory, intelligence, and capacity for work remained good in old age, as did his sight, but he became hard of hearing. His hair turned from brown to white but remained plentiful. Toward the end of his life he suffered from fainting fits. A more severe fit than usual occurred in February 1813, when his wife found him unconscious in his study. He recovered, but at the end of March had a bad cold and was feverish and out of sorts. On his own accord he decided to take a grain of emetic and vomited a considerable amount of bile. After that he fainted frequently and had pains in his arms and legs; his skin became so sensitive that he could not bear the slightest contact or the weight of his bedclothes. He died on 11 April 1813, at the age of 77. His mortal remains were brought to rest in Paris's Panthéon, in recognition of his contributions to science.

Carl Friedrich Gauss

Carl Friedrich Gauss was born on 30 April 1777 in the city of Brunswick, Germany. His father was a manual laborer who married twice. The future mathematician was the only known offspring of the second marriage. Later he tended to trace his genius to his mother, who came from a family of stonemasons. She was intelligent but apparently barely literate. Neither parent had much education.

The boy had the phenomenal memory associated with savant skills, and it was said he could calculate before he could read. At school he learned Latin, indispensable for a scholar in those days, and was strongly attracted by languages but even more by mathematics. His ability came to the notice of the local ruler, who helped him gain admission to the Brunswick Collegium Carolinum. In the library of this progressive academy he was able to study Newton's *Principia* among other scientific classics. When he was ready for university he chose the Georgia Augusta in nearby Göttingen, which had a better reputation for science than the state university of Brunswick. For three years he studied languages as well as mathematics there, largely on his own, before returning to Brunswick without taking a degree. During the next seven fruitful years he wrote his masterpiece, the *Disquisitiones arithmeticae*. He also became interested in astronomy, which offered the possibility of a career that was not available in mathematics.

In 1805 Gauss married Johanna Osthoff, the first of his two wives, whom he seems to have known from childhood. In a letter he describes her as having "the beautiful face of a Madonna, a mirror of peace of mind and health, tender, somewhat fanciful eyes, a blameless figure—this is one thing, a bright mind and an educated language—this is another, but the quiet, serene, modest and chaste soul of an angel who could not harm any creature—that is the best." She was not pleased with their rented accommodation: "shabby filthy rooms, a smoky and draughty kitchen, ancient and phlegmatic landlord and landlady" (Dunnington 1960).

When the opportunity arose to return to Göttingen as director of the new astronomical observatory, Gauss took it; he lived and worked there for the rest of his life. Soon after their marriage Johanna gave birth to a son, Joseph, who took after his father in some ways, and then to a daughter, Minna, who bore a striking resemblance to her mother. Tragically, Johanna died in 1809 after giving birth to another son, who did not survive more than a few months. The four years of marriage had made Gauss "unbelievably happy," and he was overcome with grief at her death. He wrote pages of heartrending lamentation, stained by tears.

Since Gauss felt he could not leave his young children without a mother for long, he married again within a year, this time to the daughter of a law professor at the university. The social status of his second wife was superior to his own, so he felt awkward about introducing her to his relatively humble parents. Children by her followed quickly: in 1811 and 1813 sons Eugene and Wilhelm, respectively, were born, followed in 1816 by daughter Therese.

Although Gauss's main occupation was that of director of the observatory, he was also involved in laborious projects such as the geodetic survey of the state of Hanover, which took up much of his time and energy. He spent almost six months on the survey during 1821 and not much less in each of the following four years. Living conditions in the field were often poor, sometimes miserable. He was oppressed by the heat, to which he was unusually sensitive. These were not just duties he undertook to earn a living; astronomy and geodesy were lifelong interests.

While Gauss's second marriage was not unhappy, it did not compare with the first. Minna became a chronic invalid: increasingly bedridden, she succumbed in 1831 to what was probably consumption. Unfortunately, each of his three sons was a disappointment to Gauss. Joseph acted as his father's assistant for a time, then embarked on a somewhat unsuccessful military career and ended up as a railway engineer. Both Eugene and Wilhelm emigrated to north America after extended conflicts with their father, particularly bitter in the case of the former, who seems to have inherited some of his father's gifts. Their sister Therese stayed with her father and kept house with him until his death in 1855. Gauss's own father died in 1808; nine years later his aged mother, who was nearly blind, came to end her days with her famous son.

After the death of his second wife, Gauss became increasingly introverted; the melancholy figure of later years dates from this time. When people came to see him he would talk about the days of his youth. If they started to tell him about new developments in mathematics he would often cut them short by saying he had known it all long before. When his notebooks were studied after his death it turned out there was usually much truth in this. Although he had a public reputation for being aloof, in his small inner circle Gauss appeared rather different, for he possessed the precious gift of being able to make friends with the young, who were welcome to call on him.

For a picture of Gauss in his early seventies we quote from the reminiscences of Richard Dedekind (1930):

> There were nine of us students, we all came very regularly, rarely was one of us absent, although the way to the observatory was sometimes unpleasant in winter. The auditorium, separated from Gauss's office by an ante-room, was rather small. We sat at a table, whose long sides offered a comfortable place for three, but not for four persons. Opposite the door at the upper end sat Gauss at a moderate distance from the table, and when we were all present, then the two of us who came last had to move up quite close to him and place their notebooks on their laps. Gauss wore a lightweight black cap, a rather long brown coat, grey trousers, usually he sat in a comfortable attitude, looking down, slightly stooped, with hands folded above his lap.
>
> He spoke without notes quite freely, very clearly, simply and plainly, but when he wanted to emphasize a new point in which he used an especially characteristic word, then he suddenly lifted his head, turned to one of those sitting

next to him, and gazed at him with his beautiful penetrating blue eyes. That was unforgettable. If he proceeded from an explanation of principles to the development of mathematical formulas, then he got up and in stately very upright posture he wrote on the blackboard beside him in his peculiarly beautiful handwriting; he always succeeded through economy and deliberate arrangement in making do with a rather small space. For numerical examples, on whose careful completion he placed special value, he brought along the requisite data on little slips of paper.

Gauss was a little over five feet in height, his build strong and muscular. The English diarist Thomas Hirst, who called on him towards the end of his life, recorded, "Personally he is a venerable old fellow, with a contented, manly expression. There is an extraordinary aspect of power about him and in his every word without effort he expresses the presence of manly might. He is almost eighty years of age but not a trace of superannuation about him. He can even read without spectacles" (Gardner and Wilson 1993).

Gauss was a voracious reader, owning an extensive library, including many books in English and no fewer than seventy-five volumes in Russian. He enjoyed historical works and preferred novels with a happy ending. Instances of his sense of humor are often quoted. An old friend of his wrote: "As he was in his youth, so he remained through his old age until his dying day, the unaffectedly simple Gauss. A small study, a little work table with a green cover, a standing desk painted white, a narrow sofa and, after his seventieth year, an armchair, a shaded lamp, an unheated bedroom, plain food, a dressing gown and a velvet cap, these were so becomingly all his needs" (Dunnington 1960).

Gauss was cool and reserved in judgment and had a strong sense of self-discipline. In middle age he suffered from asthma and an uncommon sensitivity to heat. He experienced a period of temporary deafness in 1838. In the evening of his life, Gauss complained about his medical and physical deterioration, but his health had been good for a man of his age until an enlarged heart led to circulatory problems. Soon after that he became housebound and able to walk only with difficulty. On 23 February 1855, at the age of 77, Gauss died in his sleep.

There are several good biographies of Gauss; two of these, Hall (1970) and Kaufmann-Bühler (1981), are listed in the references. The main problem for the biographer of Gauss is that so little is on record about his *heldenzeit,* his life before the age of 30. There are few stories of his childhood: we know why he went to study at the Georgia Augusta, but not why he left after three years. We know that on his return to Brunswick he set up house for himself, made some friends, and traveled a little. But we know far too little about what were arguably the most important years of his professional life.

Augustin-Louis Cauchy

Augustin-Louis Cauchy was born in Paris on 21 August 1789. He was christened Augustin after the month of his birth and Louis after his father, Louis-François, whose legal work for the Paris police had come to a sudden end at the outbreak of the French Revolution a few months previously. His mother came from a well-to-do bourgeois Parisian family: she had six children, of whom Augustin-Louis was the first. In 1794 the Cauchy family fled to their country house at Arcueil to escape the Terror. Arcueil was the place outside Paris where the chemist Claude Berthollet and mathematician Pierre-Simon Laplace had their estates. The boy had the benefit of meeting them and some of the other illustrious scientists who came to visit them. Lagrange, in particular, was impressed by Cauchy's ability and took an interest in his education.

We know little about Cauchy's early years except that he was a timid, frail boy with no liking for sports and games. His intellectual personality was nurtured in the intimate family circle by a very strict and pious mother and a hard-working father. In this circle he developed his exceptional capacity for hard work and the curiosity and interest in learning that, as time passed, became an almost exclusive passion for the truth.

In 1796 the Cauchy family returned to Paris, where Louis-François began to rebuild his career under the new regime. He was appointed secretary-general to the newly constituted Senate, while on Lagrange's advice his son was enrolled at the liberal École Centrale du Panthéon to study the humanities. A long-standing feature of the French educational system is the annual national competition in each subject, the Concours Général, in which the same tests are given to senior students at all the secondary schools in the country. Cauchy won the grand prize for the best student in his year. Already people were being antagonized by his austere manner and his cool exposure of his religious sentiments. According to his mother, he had many faults of character.

Cauchy entered the École Polytechnique at the age of 16 with the aim of becoming a civil engineer. At the end of two years he went on to the École des Ponts et Chaussées (bridges and highways) for two more years of study and then completed his training with field work on construction sites in the Paris region. By 1810 he was a qualified junior engineer and was sent to work on the naval base then under construction at Cherbourg. He remained there for almost three years, gaining experience as an engineer. Although his work was highly praised, he found it a strain, both physically and mentally, and his health, never robust, began to suffer. He returned to Paris on sick leave.

Among the books Cauchy had taken with him to Cherbourg to study in his spare time were Lagrange's *Traité des fonctions analytiques* and Laplace's *Mécanique céleste*. He was already making discoveries in mathematics significant enough to attract the attention of learned society in Paris. By 1812 he had perfected an important memoir on symmetric functions, which included the germ of the fundamental ideas that eventually blossomed into group theory. When his health was fully restored, Cauchy was assigned to supervise work on a construction site near Paris, but his interest in engineering had declined, and he began to look for employment where his mathematical gifts could be better engaged. An opportunity arose when Lagrange died in 1813, but although Cauchy put his name forward for the posts that thereby became vacant, he was unsuccessful. However, the situation was transformed after the defeat of the French at Waterloo, the second abdication of Bonaparte, and the return of Louis XVIII. Cauchy's father managed to retain his high position in the Senate, now renamed the Chamber of Peers.

The Restoration was the most fruitful period of Cauchy's career. He had always been strongly royalist, and new opportunities opened up for him as scientists too closely identified with the Revolution were replaced. In 1816, after several previous attempts, Cauchy was elected a member of the Paris Academy. About the same time, he was appointed associate professor of analysis at the École Polytechnique, and before long, after a politically motivated reorganization had produced a few vacancies, he was promoted to full professor in analysis and mechanics.

Cauchy taught at the Collège de France and at the university as well as at the École Polytechnique. Often in his lectures he introduced new ideas and more rigorous methods, such as the modern definition of continuity. He made efforts to reform the syllabus at the École Polytechnique, which were not altogether successful; there were objections that his courses were over-ambitious and that he gave too much time to pure rather than applied mathematics. After all, it was said, the Polytechnique had been founded to train engineers, not mathematicians.

A conscientious teacher, he made a good impression on at least some of his students: "We all found that [Cauchy] was extremely energetic, good-natured and tireless. I often heard him repeat and review, for several hours on end, whole lessons that we had not understood clearly," wrote one of his former students eulogistically after Cauchy's death. "His love of teaching, which bordered on pure zeal, brought with it a kindness, a simplicity and warmth of heart that he retained until the end of his life" (James 2002). However, others objected to the way his courses ran well over the allotted time, and occasionally there were expressions of hostility to a teacher who did not attempt to conceal his royalist views when addressing an audience of mainly liberal students.

By 1818, Cauchy was ready to get married and establish a home of his own. His bride, Aloise de Bure, was a member of a solid old bourgeois family of booksellers and publishers. The marriage, at the fashionable Paris church of Saint Sulpice, was a grand affair, attended by academic, political, and religious notables; the king and the entire royal family displayed their goodwill and esteem by adding their signatures to the marriage contract. Essentially a marriage of convenience, it seemed to work well, at least at first. The de Bures, who were also Cauchy's publishers, provided the young couple with an apartment in their Paris mansion and enabled them to spend their summers at a family property in Sceaux, just beyond Arcueil. Aloise gave birth to two daughters.

In addition to his professional work, Cauchy devoted much of his time and energy to the Congrégation de la Sainte Vierge, an organization of "young Catholics of good families" aimed at combating faithlessness, irre-ligion, and secularism. He had joined this organization when a student at the École Polytechnique, where it was particularly active. In the Academy and elsewhere he was one of the champions of clericalism. He detested liberalism in all its forms, but he particularly detested liberal Catholicism. Those who came to see him about mathematics often found that what he wanted was to convert them to his brand of religion.

Cauchy's tastes and ways were austere in the extreme. He was a solitary person who pushed himself excessively. His study was small and modestly furnished. Most often he did his writing at a small desk—a table really, without pigeonholes—which in the evenings was lit by two simple wax candles with shades. Everything in this room and in his library was in perfect order. He never went out without extinguishing the fire in the fireplace.

When the ultraconservative Charles X succeeded Louis XVIII in 1824, his policies were in accordance with Cauchy's reactionary views. After Parliament was summoned and almost immediately dissolved, the king began to rule by decree. Rioting began, and Paris was soon in the grip of insurrection. Meanwhile, Cauchy's health deteriorated due to overwork combined with dismay, even fury, at the return to power of the liberals. When the unpopular king made way for his grandson, the juvenile Duke of Bordeaux, and fled to Scotland, Cauchy left for Switzerland on what began as sick leave but turned into an eight-year self-imposed exile, a personal demonstration of loyalty to the Bourbon monarchy. As an absentee, he was dismissed from his posts, leaving only his membership in the Academy.

Throughout the first three years of this period, Cauchy mainly based himself in Turin, where he was appointed professor of mathematical physics, quickly learned Italian, and lectured in that language. Someone who attended one of his courses described them as "very confused. He skipped suddenly from one subject to another, from one formula to the next, with no attempt to give a connection between them. His presentations were obscure clouds, illuminated from time to time by flashes of pure genius" (Belhoste 1991). In 1833 he moved to Prague, where the exiled king had by then established himself, and was appointed tutor to the Duke of Bordeaux. Cauchy tried hard to interest his juvenile pupil in elementary science but was treated with contempt. Cauchy's family, who had hitherto remained in Paris, came out to join him in Prague.

The Orleanist Louis-Philippe, who succeeded Charles X, was too liberal for Cauchy. When he in turn fell from power in 1848, Cauchy returned from exile, hoping for a Bourbon restoration. But after an interregnum, the throne passed to Napoleon III. Cauchy was able to get reinstated at the Sorbonne, where he resumed lecturing until the introduction of an oath of allegiance caused him to give up his professorship. Efforts by his supporters to reinstate him at the Bureau des Longitudes and the Collège de France were unsuccessful.

Cauchy's years of exile were a period of low productivity compared with what came before. The rising stars of the new generation of French mathematicians were deprived of Cauchy's teaching, to their great loss. Nevertheless, his influence on the younger generation of mathematicians remained important, and he continued to produce research. In later years, much of his time was spent on charitable work in Sceaux, to which he devoted his whole stipend. Academic disputes, usually with political overtones, took up much of his time as well. The machinations of the academicians, in which Cauchy played an all-too-active part, were a continual drain on his energy. He also became involved in fruitless arguments about priority of discovery. His health, never robust, began to give cause for concern. On medical advice, Cauchy left Paris for Sceaux on 12 May 1857, intending to spend the summer there, but his health suddenly deteriorated and he died eleven days later, at the age of 67.

A vast collection of Cauchy's notebooks and other manuscript material, including a great deal of poetry in both French and Latin, was wantonly destroyed just before the Second World War. In his lifetime, he published a huge amount of mathematical work, both research monographs and textbooks, as well as a steady stream of research papers. On his instructions some further unpublished works were added to this posthumously to make up his voluminous *Complete Works*. Unfortunately, much of what he wrote is chaotic; a result is stated, then refuted, only to be stated once more; a procedure is severely criticized, only to be applied successfully at the next opportunity; for no apparent reason notations are changed back and forth. He was the opposite of Gauss, who published so much less than he might have. What Gauss did publish was profound, but few could understand it. Cauchy's works seem superficial by comparison, but they soon stimulated further research. Sometimes he forgot about something he had written earlier and published the same paper twice. His mathematical style was abstract and rigorous, with a tendency toward what some saw as willful obscurity.

If Cauchy hit on a new idea he could not bear to wait before publishing it. The weekly *Comptes rendus* of the Academy was not founded until 1835. Cauchy started his own monthly journal, *Exercices de mathématiques*, in 1825. Its pages were filled exclusively by Cauchy himself, with the most improbable choice of topics in the most improbable order. Five volumes of the *Exercices* appeared before he left Paris. In Turin he revived this under-

taking—publishing even in the local newspaper—and continued it in
Prague and back in Paris, finally completing a total of ten volumes. He
published in other journals too, and put out at least eighteen memoirs as
well as many textbooks. Sometimes his activity seems explosive even by his
own standards. At three meetings of the Academy in August 1848, he submitted no fewer than five notes and five memoirs. At nine meetings soon
afterwards he submitted nineteen notes and ten memoirs. Overwhelmed by
this deluge, the Academy made a rule, still in force today, that papers
printed in the *Comptes rendus* should not exceed four pages in length.
Someone who attended one of Cauchy's undergraduate lectures about this
time found it a strange experience. Apparently he spent a whole hour
computing the square root of seventeen to ten decimal places by a method
he had just invented, but which was well known to all of the students.

Self-absorbed, self-righteous, and sanctimonious, no wonder Cauchy
was unpopular. The opinion of a contemporary was typical: "His harsh and
rigid spirit, his lack of indulgence towards the young who follow a scientific
career, make him one of the least likeable of the savants and certainly one of
the least liked." Had it not been for his intolerant, almost fanatical, political
opinions, he might have achieved much more: as it was he wasted some of
his prodigious gifts and occasionally misused the power they gave him.

Cauchy seems to have gone out of his way to pass judgment on the work
of others. As he became more senior and more influential, the aspiring
young who looked to him for encouragement and support were often disappointed. However, it would be wrong to think that Cauchy never recognized the merits of the research of others. When he refereed a paper, he
reported on its merits honestly, even when it overlapped with his own work.
Of all the mathematicians of his period, he was perhaps the most careful in
quoting others, and if he was shown to be in error, he candidly admitted it.
Self-absorbed he may have been, but he never allowed himself to be swayed
by self-interest.

Cauchy was a proud man, a man of passionately held convictions who,
when occasions arose for him to defend or explain those things that he
regarded as the truth, consistently refused to allow himself to be affected by
selfish considerations of personal convenience. For example, in the political
sphere, his adherence to the Bourbon cause was, for better or for worse,
absolute and unyielding; Cauchy chose exile over perjury. His belief in
Catholicism was uncompromising. After a lecture on natural history at the

academy, he rose to protest: "Even if these things would be as true as I think, they are wrong. It would not be convenient to disclose them to the public, given the devilish state into which our misbegotten Revolution has hurled public opinion. Any such talk can only harm our holy religion."

Feeling uncomfortable in his own era and often misunderstood by his contemporaries, Cauchy found refuge in mathematics. He displayed a stubbornness and rigidity of character that his contemporaries frequently mistook for narrow-mindedness, particularly in political matters. But without this stubbornness it is hardly likely that he could have persisted in attacking and solving so many difficult and fundamental mathematical problems, some of which did not yield to his efforts for long periods of time.

What might have been the nature of the major illness that first afflicted Cauchy in his early twenties and recurred later in his life? The symptoms, we are told, were depression with moments of excitement and physical weakness. When his mother came to fetch him from Cherbourg, she found him in a weakened and depressed condition, while his father said that his son's health was profoundly altered by overwork. The illness seems to have been as much emotional as physical, with pronounced psychological overtones. An ill-defined sickness, it lasted on and off for about a year on the first occasion. The long period of convalescence was periodically interrupted by a few weeks in which Cauchy was able to work. Numerous other facts point to the conclusion that, throughout his life, Cauchy suffered from episodes of depression.

He also displayed some traits of Asperger syndrome. He was a timid, frail boy with no liking for sports or games, and he had an exceptional capacity for hard work and a curiosity and interest in learning. He was formal, too cold, and tactless. He engaged in a marriage of convenience. His study was small and modestly furnished, his library in perfect order. In his personal life, as in his work, he was both volatile and stubborn. His was a cold, aloof, inflexible personality, characteristic of the autistic superego. Cauchy did not so much master mathematics, it has been said, mathematics mastered Cauchy.

Lejeune Dirichlet

The mathematicians profiled so far were not remarkable for their good nature, so far as we know. We now turn to one who most certainly was.

Peter Gustav Lejeune Dirichlet was born on 13 February 1805 in the small town of Düren, midway between Aachen and Cologne. This part of the German-speaking Rhineland was then within the French empire. Dirichlet's father, the local postmaster, was said to be gentle, pleasant, and amiable; his mother cultivated and intellectual. His parents did their best to give their clever son a good education, although this cannot have been easy since they had ten other children to provide for and were not at all well off.

At school the boy showed a precocious interest in mathematics, using his pocket money to buy mathematical books. When he was twelve his parents sent him to the Gymnasium in Bonn, where he is reported to have been an unusually attentive and well-behaved pupil who showed a particular interest in recent French history as well as mathematics. After two years he went to the excellent Jesuit college in Cologne, where he was fortunate to have the physicist Georg Simon Ohm among his teachers.

At the early age of sixteen, Dirichlet was ready for university, where his parents wanted him to study law. Although not unwilling to do so, he preferred to study mathematics, and persuaded his parents to let him. At that time, the general standard of German universities in mathematical studies was mediocre. There were no outstanding German mathematicians apart from Gauss, whose time and energy were mainly taken up with astronomy. Dirichlet decided it would be best to pursue his studies in Paris, where there was a galaxy of talent, including Augustin Cauchy, Joseph Fourier, Pierre-Simon Laplace, Adrien-Marie Legendre, and Simeon-Dennis Poisson. Shortly after he arrived in the French capital, at the age of seventeen, he caught smallpox, but this did not interrupt his studies for long. Although he suffered from poverty, he always looked back on this first year in Paris with intense pleasure.

The next year Dirichlet had an unexpected stroke of good fortune. In the summer of 1823 General Maximilien Foy, a national hero of the Napoleonic wars and then the leader of the liberal opposition in the Chamber of Deputies, was looking for a house tutor to teach his children the German language and literature. Dirichlet secured this well-paid and pleasant post, in which he was treated as a member of the family. Madame Foy, who praised him highly, described him sitting on a little stove all day long, teaching her children and doing his own work at the same time. He also improved her knowledge of German, while she corrected Germanisms in his French. Through the Foys, Dirichlet had the opportunity to meet many of the scientists then living in Paris. Two who took a special interest in the young man were Joseph Fourier, permanent secretary of the Academy, and the able and influential Alexander von Humboldt. He also became friendly with the young Norwegian genius Niels Abel, who was in Paris during the second half of 1826.

Dirichlet always considered himself German rather than French, although his father was of French origin. Much as he enjoyed the brilliant intellectual life of Paris, his aim was to find a position in Germany. To secure a university post, Dirichlet first needed to obtain the *venia docendi,* which in turn required a degree from a German university. On the recommendation of Gauss and Humboldt, the University of Cologne was persuaded to award him an honorary doctorate, after which he was able to habilitate at the University of Breslau, where he was appointed first *privatdozent* and then associate professor. A disciple of Gauss, he wrote about theory of numbers at this stage of his career.

After Paris, Dirichlet found the conditions for scientific work at the German university uninspiring. He was not popular with his colleagues, and the students did not appreciate his teaching style. The mathematicians in Paris, particularly Fourier, urged him to come back, promising that he would be offered a position at the Academy, but Dirichlet preferred to remain in Germany and decided to try for a position in Berlin. In 1827 he wrote to his mother, "There is so little here to offer from the scientific side that I will mobilize all the forces I can to bring about my transfer to Berlin" (Kümmel 1889–1897). Again he sought the support of Gauss, who replied: "It is all the more pleasing to me that you have a great attachment to that part of mathematics [the theory of numbers] that has always been my favourite field of study, however seldom I have pursued it. I dearly wish you in a situation in which you will have as much control over your time and the choice of your work as possible. Immediately after the appearance of my *Disquisitiones,* I myself was very much hindered by other business, and later by my external circumstances, from following my inclinations to the degree that I would have wished" (Kümmel 1889–1897).

The unfortunate Abel also hoped for a position in Berlin, and, as his biographer Arild Stubhaug (2000) points out, Dirichlet might have unwittingly spoiled his chances. There was a scheme to establish a Technische Hochschule (technical university) in the Prussian capital, modeled on the École Polytechnique in Paris, and if this had succeeded there might have been a place for both of them. As it was, Abel returned to Norway in 1827; the next year Dirichlet succeeded in moving to Berlin, initially as a teacher in the military academy, a post he retained for almost thirty years. In 1831 he was elected to the Berlin Academy, which had a strong mathematical tradition going back to Leonhard Euler and Joseph-Louis Lagrange. The same year he was appointed associate professor at the university; he was finally promoted to full professor in 1839, although denied the full status of *ordinarius* because he had not written a dissertation in Latin.

Soon after Dirichlet arrived in the Prussian capital, Alexander von Humboldt introduced him to members of the prominent family of Mendelssohn, descendants of Moses Mendelssohn, who prepared the ground for the Jewish emancipation movement. (They added the name Bartoldy when, like many other Jews, they converted to Christianity for reasons of expediency.) One of them was the celebrated composer Felix, whose elder sister Fanny described Dirichlet as "a very handsome and amiable man, as full of fun and spirits as a student, and very learned" (Hensel 1881). She was engaged to marry the artist Wilhelm Hensel, but her younger sister Rebecca was unmarried and within a year became Dirichlet's wife. The young couple began their married life in the home of Rebecca's parents, where their first child Walther was born; another son did not survive infancy, but a third son, Ernst, reached maturity. Later Dirichlet's protégé Ernst Kummer would marry Ottolie Mendelssohn, a cousin of Rebecca's.

At Berlin Dirichlet succeeded in modernizing mathematical studies by improving examinations, introducing new methods of teaching, and developing the curriculum until the university could offer a mathematical training that rivaled that provided by France's École Polytechnique. He conducted an informal seminar at which the more able students lectured on recent research, an innovation he would later introduce at the Georgia Augusta. Through his lectures, through his many students, and through the series of scientific papers of the highest quality that he wrote and published during this period, Dirichlet strongly influenced the development of mathematics in Germany and elsewhere during the twenty-seven years spent as a professor in Berlin. As Jacobi stated: "Dirichlet alone, not I, nor Cauchy, nor Gauss knows what a completely rigorous proof is. Rather we learn it first from him. When Gauss says he has proved something it is very clear; when Cauchy says it, one can wager as much pro as con; when Dirichlet says it, it is certain" (James 2002). He practiced the art of overcoming problems with a minimum of blind calculation and a maximum of perception.

When the English diarist T. A. Hirst was in Berlin in 1852, he called on the 47-year-old Dirichlet and found him:

> a rather tall, lanky-looking man, with moustache and beard about to turn grey—with a somewhat harsh voice and rather deaf; it was early, he was unwashed and unshaved, with his "schlafrock" slippers, cup of coffee and cigar. I thought, as we sat each at the end of the sofa, and the smoke from our cigars carried question and answer to and fro, and intersected in graceful curves before it rose to the ceiling and mixed with the common atmospheric air: if all be well, we will smoke our friendly cigars together many a time yet, good-natured Lejeune Dirichlet. (Gardner and Wilson 1993)

The two men got on well, and Hirst was invited to spend social evenings with the Dirichlets and their friends.

Dirichlet was said to be an excellent teacher. Hirst wrote of his lectures:

> Dirichlet cannot be surpassed for richness of material and clear insight into it. As a speaker he has no natural advantages—there is nothing like fluency about him, and yet a clear eye and understanding make it dispensable. Without an effort you would not notice his hesitating speech. What is peculiar in him, he never sees his audience—when he does not use the blackboard, at which time his back is turned to us, he sits at the high desk facing us, puts his spectacles on his forehead, leans his head on both hands, and keeps his eyes, when not completely covered with his hands, mostly shut. He uses no notes, inside his hands he sees an imaginary calculation, and reads it out to us—that we understand it as well as if we too saw it. I like that kind of lecturing. (Gardner and Wilson 1993)

In 1854 Dirichlet attended the celebration of Gauss's academic jubilee at the Georgia Augusta. At one point he noticed Gauss trying to light his pipe with a piece of the original manuscript of the *Disquisitiones*. Horrified by this act of sacrilege, he rescued the paper and treasured it ever afterwards. Gauss died the following year, and the Georgia Augusta was anxious to find a successor of great distinction. The choice fell on Dirichlet, who responded that he would accept unless he was able to obtain relief from the drudgery of his work at the military academy, where the cadets were becoming increasingly unpleasant to him. The Prussian ministry of education did not seem to take the situation very seriously, and it was only after it was too late that he was offered some improvement in his teaching load and salary.

So the Dirichlets left Berlin in the autumn of 1855 and settled at Göttingen, where he occupied the post left vacant by Gauss. When Rebecca's widowed sister Fanny died they adopted her orphaned son. Dirichlet purchased a pleasantly situated house with a garden. In the words of Minkowski (1917): "The peace of a smaller town, something [Dirichlet] had not enjoyed since his youth, was sufficient compensation for the superficial convenience of living in a large town like Berlin. True he had fewer students than in Berlin . . . [but he] attracted many young mathematicians to Göttingen." At the Georgia Augusta, Dirichlet found a circle of congenial colleagues, especially Richard Dedekind. They were in contact almost daily: "I owe him infinitely much," Dedekind wrote to his sisters. Dirichlet also had a number of excellent students, including Gotthold Eisenstein, Leopold Kronecker, and Rudolf Lipschitz, enjoying an intellectual and appreciative audience for his lectures.

The rising star in the mathematical firmament at the time was Bernhard Riemann, whose profile is given in this book. The young man had attended Dirichlet's lectures when studying in Berlin and then followed him to the Georgia Augusta. Riemann sought Dirichlet's advice on his habilitation thesis, writing of Dirichlet: "He went through my dissertation with me." Riemann thought Dirichlet the greatest living mathematician after Gauss. According to Klein (1926/1927), "Riemann was bound to Dirichlet by the strong inner sympathy of a like mode of thought. Dirichlet loved to make things clear to himself in an intuitive substrate; along with this he would give acute logical analyses of foundational questions and would avoid long computations as much as possible. His manner suited Riemann, who adopted it and worked according to Dirichlet's methods." Dirichlet also helped the impoverished *privatdozent* in practical ways, for example, by finding him a minor position at the university, which brought in some extra income and entitled him to a small rent-free apartment.

Early in 1858, Dirichlet devoted himself to preparing a memorial address in honor of Gauss. That summer at Montreux, on the lake of Geneva, he suffered a severe heart attack, after which he was able to return home to Göttingen only with great difficulty. While Rebecca was nursing him back to health, she suffered a fatal stroke. Dirichlet's condition steadily deteriorated, and he died on 5 May 1859, at the age of 54.

Both directly and indirectly, through his many disciples, Dirichlet was mainly responsible for the transmission to the German-speaking countries of the French discipline of mathematical physics, which he had learned at first hand in Paris. As Butzer (1987) has shown, practically all the great German theoretical physicists of the nineteenth century were strongly influenced either by Dirichlet himself or by his students or their students.

Chapter 5 } Hamilton, Galois, Byron, and Riemann

William Rowan Hamilton

Archibald Hamilton, a lawyer who acted for the ardent Irish nationalist Archibald Hamilton Rowan, lived in Dublin. Like his wife, Sarah Hutton, he was of Scottish ancestry. At the stroke of midnight on 4 August 1805, their only son William, the future mathematician, was born. Possibly because his father was short on money, the boy was sent at the age of 3 to live with his uncle, the Reverend James Hamilton, in the sleepy village of Trim, forty miles northwest of Dublin. The uncle, a classical scholar who had graduated from Trinity College Dublin, was headmaster of a school in Trim. William had little contact with his parents during childhood; his letters to his father were formal and stilted. Occasionally his four sisters joined him at Trim, and on both sides of the family there were relatives who took an interest in his progress. When his mother died in 1817, his father remarried, but died shortly afterwards.

The boy began his education as soon as he arrived in Trim, quickly revealing himself to be a child prodigy. His uncle aimed to get him into Trinity, and gave him every encouragement. Languages came easily to Hamilton, and at the age of 10 he was said to know not only the classical languages of Latin, Greek, and Hebrew, but also a little of some modern European languages and a smattering of various oriental languages (apparently because his father thought he might make a career in the East India Company). He was also something of a calculating prodigy and retained an uncanny ability in this respect throughout his life.

Hamilton's serious mathematical education began about the age of 13 with the study of Euclid, and he went on to read the works of Newton and Laplace. In 1824, at 19, he entered Trinity College Dublin and that same year published a note correcting a minor error in Laplace's work. Luckily the mathematics curriculum at the college had just been reformed and was based to a considerable extent on what was taught at the École Polytechnique. During his four years at Trinity, Hamilton demolished the competition in one examination after another. The college awarded him two separate *optimes* for his performances: so rare was the honor that no optimes at all had been awarded for twenty years.

While still a student, Hamilton wrote what he planned would be the first part of a treatise on optics. The paper, which he presented to the Royal Irish Academy, dealt with caustics, the patterns of light produced by reflection and refraction. At first it was turned down, but the following year a revised version was accepted and published by the Academy under the title "The Theory of Systems of Rays." This made his reputation and set the course of his research for the next decade.

At this point, Hamilton might have taken the fellowship examination, as had been his ambition since he first came to the college. However, another possibility arose. In 1827 the chair of astronomy at Trinity became vacant, and Hamilton was invited to apply. In spite of his age and lack of experience, Hamilton was appointed Andrews Professor of Astronomy and consequently became *ex officio* astronomer royal of Ireland and director of the Observatory at Dunsink, although he had yet to take his degree. Hamilton retained these latter positions throughout his life and made the observatory his home, but he was never a success as a practical astronomer and after a few years gave up such work to concentrate on mathematics.

The most important of Hamilton's major mathematical works is *The General Method in Dynamics;* it is also the one on which he spent the least time. It exhibited to an extraordinary degree his ability to create rapidly an enormous body of theory almost unparalleled in generality and abstraction. It grew directly out of his theory of systems of rays, which in turn grew out of Lagrange's analytical mechanics. His work soon became known on the Continent, where Jacobi referred to Hamilton as "the illustrious Astronomer Royal of Dublin," and as "the Lagrange of your country." With the advent of the quantum theory in the twentieth century. Hamilton's Principle came into its own, because it was the one form of classical mechanics that could be transferred almost directly to quantum mechanics.

Hamilton's optical researches, based on the undulatory theory of light, culminated in his prediction of the phenomenon of conical refraction in biaxial crystals, an idea that was later proved experimentally. The German mathematician and physicist Plücker wrote: "No experiment of physics has ever made such an impression on me . . . it was a thing unheard of and completely without analogy" (James 2002). In 1835 this discovery won the 30-year-old Hamilton a Royal Medal from the Royal Society of London and a knighthood from the Lord Lieutenant of Ireland, on behalf of the king. In many ways this was the zenith of his professional career.

Hamilton had been christened William but later added Rowan, the surname of his father's former employer, in the expectation of patronage from that quarter; thus in later years he was known as Sir William Rowan Hamilton. During the 1830s he led an extremely busy life. He seemed to have almost boundless physical energy, and we read of him walking along the parapet of the observatory roof and other escapades. He was an active member of the British Association for the Advancement of Science, delivering papers, chairing sections, and making speeches at the banquets that accompanied each meeting. In his prime he was a convivial and jovial man, although socially naive and eccentric. He was elected president of the Royal Irish Academy. During this decade he had various romantic attachments, but one in particular stands out. This was to Catherine Disney, with whom he became infatuated. Although she was strongly attracted to him, family pressure caused her to marry someone else, and as a result Hamilton contemplated suicide.

After another young lady rejected his advances, he began to feel that marriage was going to elude him, and in desperation proposed to Helen Maria Bayly. She came from a family of minor gentry from County Tipperary, where she lived on a farm with her widowed mother. A friend said that she had no "striking beauty of face or force of intellect," "could not manage well the concerns of a household," and was totally unable "to exercise a controlling influence over the habits of her husband" (Hankins 1980).

Helen married Hamilton with reluctance and after much vacillation in April 1833; after the marriage she continued to spend lengthy periods with various members of the Bayly family. Their first child, William Edwin, was born in May the following year. Hamilton recited Wordsworth's "Ode on Intimations of Mortality" to the baby. In later life, William was to create financial problems for his father. A second son, Archibald Henry, was born a year later; he grew up to become a clergyman. Helen and William's only daughter Helen was born in August 1840. All three children were regarded as rather eccentric (Hankins 1980).

Hamilton was a very religious man, and his wife's strong piety was important to him. Before their marriage, he recognized that she was likely to be chronically ill and that she was pathologically shy and timid. In spite of this she remained the central figure in his life, and to some extent drew him into her own habit of seclusion. If he felt her weakness of health and character as a burden, it was one that he willingly accepted. Yet his friends noticed a change in him. He became more solitary, more introspective, and much more defensive than before. He tended to dwell on past accomplishments rather than show enthusiasm for new pursuits, and he gradually began to drink more than was good for him.

Hamilton regarded himself as a good lecturer to popular audiences and first-year students. He did indeed have a flair for oratory, but none at all for elementary instruction in mathematics. Every such attempt began with explanations so elementary and obvious that they were embarrassing to his listeners; then suddenly he would switch to material totally incomprehensible to them. Nor was he any better at written instruction; his papers constantly digress.

Hamilton was considered by some of his contemporaries to approach Isaac Newton in intellectual power. For various reasons his influence was not as great as it might have been; in particular he did not try to establish an Irish school of mathematics. Because he worked at the observatory outside Dublin, mainly on his own, his contact with students was minimal.

In the last part of his life Hamilton indulged in romantic fantasies and conducted a prolonged and secretive correspondence with Catherine Disney, with whom (or with an idealized version of whom) he never ceased to be infatuated. In the summer of 1865, having suffered from occasional bouts of gout over the years, Hamilton became seriously ill. The greatest problem for those around him was keeping him from worrying about work and money. His more or less permanent bank overdraft was increasing alarmingly. By the end of August it became clear that he would not live much longer, and death came on 2 September 1865, in his sixty-first year. Hamilton's doctor was at pains to report that, although the cause of death was uncertain, it was not alcoholism; however, there can be no doubt that this was a serious problem in the later part of his life. Walker and Fitzgerald

(2006) provide much more information about Hamilton's personality and conclude that "his truly exceptional mind, eccentricity, and creative nature all point to him being an Asperger genius."

Evariste Galois

Evariste Galois was born in the town of Bourg-la-Reine, on the outskirts of Paris, on 25 October 1811. He was the second son of Nicholas-Gabriel Galois, the director of a boarding school and later mayor of Bourg-la-Reine, and Adelaide-Marie (née Demante), who came from a family of lawyers. Both parents were well educated, the father said to have been imaginative and liberal, the mother headstrong and eccentric. Evariste had a happy, if unconventional, childhood. Like Gauss, he seems to have had a phenomenal memory. Up to the age of 12 he was educated entirely by his mother, who instilled in him a knowledge of the classics and a skeptical attitude towards religion. She was an intelligent, lively, and generous woman, of strong character, but later seems to have been quite unable to restrain her brilliant, sensitive, but willful son.

In 1823 the future mathematician began his formal education by entering, as a boarder, the grim but prestigious Lycée Louis-le-Grand. Almost as soon as he arrived there was something of a revolt by a number of the boys, who suspected the authorities of planning to reintroduce the conservative Jesuits to the staff; the outcome was that forty boys were expelled. At first Galois did well in examinations, coming out near the top in the national Concours Général. Interest in mathematics does not appear to have run in the Galois family, and Everiste did not begin studying it at the Lycée until his fourth year there, when he made phenomenal progress in the subject at the expense of the rest of the curriculum.

This sudden obsession with mathematics seems to have distorted his psychological development. His mother felt that his character changed from being serious, open, and loving to taciturn and peculiar. Comments about his character being "singular" and "closed" but "original" began to appear in his school report; the authorities accused him of putting on affected airs and being eccentric and ambitious. It was said that he was bizarre, that there was something furtive about his character, that he teased his fellow pupils.

Soon Galois believed he was ready to enter the École Polytechnique but, due to his lack of preparation in the standard syllabus, failed the difficult entrance examination. Fortunately he then came to the notice of a perceptive mathematics teacher named Louis-Paul-Emile Richard, who recognized his exceptional gifts. Although still only 17, Galois was already working on his theory of equations and submitted his first paper on the subject to the Paris Academy two months later. Cauchy, who was referee, seemed favorably impressed and said he would present it to the Academy himself. Unfortunately, Cauchy was unwell on the day he was to have made the presentation. Soon afterward Galois (perhaps on Cauchy's advice) combined his researches into a memoir to be submitted for the Grand Prize in mathematics.

Galois's father was an ardent Bonapartist at a time when royalists were in the ascendant, and in July 1829, he committed suicide. The immediate cause was a spiteful campaign against him by some local royalists, led by the parish priest, who were trying to force him out of office. The mayor was very popular locally, and the entire population of Bourg-la-Reine contributed to the cost of a memorial. His son was grief-stricken; this was a turning point in his life. A few days after his father's death he again failed the entrance examination for the École Polytechnique (only two attempts were allowed). He then tried for the somewhat less prestigious École Préparatoire, the institution that later became the École Normale Supérieure. Galois was successful in meeting the entrance requirements and was accepted in February 1830.

Shortly afterward he learned of the death of Niels Abel, and found that Abel had already obtained some of the results that he had just discovered himself. Galois then published several short papers containing most of the work on the theory of equations that we know as Galois theory. He also submitted his memoir on the theory to the Academy for the Grand Prize, but Fourier, the permanent secretary concerned, was seriously ill and died shortly afterward. Apparently he had taken the memoir home to read, and it was never returned to the Academy. Meanwhile, Galois passed the first-year examination at the École Préparatoire, but only in fourth place out of eight candidates. He was becoming increasingly involved in political activity.

As touched upon in the profile of Cauchy, during the summer of 1830 mounting protests against the rule of the ultraconservative King Charles X led to an insurrection, beginning with rioting on the streets of Paris. Students from the École Polytechnique, at that time a center for republicanism, participated in this, but those at the École Préparatoire, where the director was a staunch royalist, could not get out onto the streets to join them because the gates of their building had been locked. At the end of the "three glorious days," in which thousands died, the rebels were masters of the capital and the king, having lost control, abdicated and fled. However, the rebels, consisting mainly of Bonapartists and republicans, were unable to prevent the more moderate, but still conservative, Duke of Orleans from accepting the crown as King Louis-Philippe.

Republicans like Galois were bitterly disappointed with the outcome. The restoration of monarchy gave him an ideal focus for his resentment over his father's death. When he returned to Bourg-la-Reine for the summer vacation his family was amazed at the transformation in his character: he had become a republican firebrand. Galois is thought to have joined a group of republican activists called the Société des Amis du Peuple: before long this was suppressed and its more militant members joined a republican stronghold called the Artillery of the National Guard. This was an armed organization not under the control of the regular army, and not surprisingly it was disbanded before the end of the year.

Meanwhile Galois was expelled from the École Préparatoire, which had become the École Normale Supérieure. The reason given for his expulsion was the publication of an article attributed to him calling for the students to be armed. He had been busy with political agitation among his fellow students, but they were at most lukewarm in their support. Because of his expulsion he lost the grant he had been receiving. He tried to earn a living by teaching, giving tuition in mathematics to bored schoolboys. A lecture course he gave on his own theories was not a success. He also attended lectures at the Academy, where he displayed a capacity for rudeness.

After the Artillery Guard was disbanded, nineteen of its members refused to hand over their arms. They were put on trial in April 1831, and when they were acquitted, a celebratory banquet was organized at a restaurant. The novelist Alexandre Dumas, who was present, wrote afterwards that "it would be difficult to find two hundred people in the whole of Paris who were more hostile to the government" (Rigatelli 1996). Some of them, including Galois, illegally wore the uniform of the Artillery Guard. Although the organizers were anxious to avoid trouble with the police, matters got out of hand when Galois rose and proposed an ironic toast to the new king, with a dagger in his hand. Others followed suit and the banquet ended in turmoil. Galois was arrested the following day and held in prison until June. He was then tried for threatening the life of Louis-Philippe but was acquitted almost at once on the grounds that he was young and foolish.

Galois was arrested again on Bastille Day 1831 for illegal possession of weapons (a loaded rifle and several pistols as well as the dagger) and for wearing the uniform of the illegal Artillery Guard. When the police came for him he was out on the streets at the head of a procession. He was held in prison until October and then sentenced to a further six months, confirmed on appeal. He became very despondent and once, thinking of his beloved father, attempted suicide. The following March, he was placed on parole and transferred to a nursing home as a measure against cholera, which was then epidemic in Paris. In those relatively pleasant surroundings, Galois started to think about mathematics again and managed to write a few philosophical essays.

Meanwhile Galois was waiting for some news about his memoir on the theory of equations, part of the work he had submitted for the Grand Prize. Eventually, while back in prison, he heard from the Academy that the referee had advised against its acceptance but recommended encouraging the author to submit a more complete account of his theory. Galois reacted angrily to this and resolved to have no more to do with the Academicians. He did in fact revise his work (the original version has disappeared), with private publication in mind, but poured most of his energy into the five-page preface, a scorching condemnation of the scientific establishment and particularly the Academicians, "who already have the death of Abel on their consciences."

Galois was released on 29 April 1832. Frustratingly little is known about the next, and final, month of his life. The police records, which might have thrown some light on it, were destroyed in the Paris Commune of 1871. On 25 May he wrote to Chevalier, a close friend of his from the École Normale, expressing his complete disenchantment with life and hinting that a broken love affair was the reason. The woman in question was the daughter of the resident physician at the nursing home. Copies made by Galois of two letters he received from her exist, although they are incomplete and only partly readable. One, dated 14 May, says "Please let us break up this affair." The other mentions sorrows that someone else had caused her in such a way that Galois might have felt obliged to come to her defense.

On 30 May, Galois fought a duel with pistols. There is some doubt about the identity of his opponent, but it seems to have been a prominent republican activist named Pescheux d'Herbinville, one of the nineteen tried the previous year. Galois was shot in the abdomen and lay unattended for hours until a passerby took him to hospital. He died of peritonitis the following day, at the age of 20, with his brother Alfred beside him, and was buried in a common grave somewhere in Montparnasse. Given his mental state, there may be no rational explanation for the fatal duel, but no doubt the suicide of his beloved father and his own self-destructive tendencies were among the factors.

Galois was convinced that he was going to die over something small and contemptible; the tragedy is that he let it happen. Perhaps the broken love affair had something to do with it, but there are indications that he already believed he was doomed much earlier. When he first went to prison in 1831, one of his comrades wrote a letter quoting Galois as follows: "And I tell you I will die in a duel over some low-class coquette. Why? Because she will invite me to avenge her honour which another has compromised." In letters he wrote to friends the night before the duel, Galois again referred to an "infamous coquette." He added, "Forgive those who kill me for they are of good faith."

Various imaginative reconstructions have been built up around these strange words. One recent theory, which seems more plausible than most, is that Galois, wishing to be a martyr to the republican cause, had arranged a false duel. He meant to be killed, with the idea that this would be blamed on the police, and there would be a riot in protest. To avoid his death being blamed on his political friends, he laid a false trail of messages—the letters and other writings hardly make sense if he was going to fight a normal duel. If there is anything in this theory, his sacrifice was in vain, because his death caused no riots. The night before he died Galois wrote at length to his friend Auguste Chevalier, also a mathematician, outlining his discoveries:

> I have often in my life ventured to advance propositions of which I was uncertain; but all that I have written here has been in my head nearly a year, and it is too much to my interest not to deceive myself that I have been suspected of announcing theorems of which I had not a complete demonstration . . . Make a public request to Jacobi or Gauss to give their opinions not as to the truth but as to the importance of these theories. After that I hope some men will find it profitable to sort out this mess. (Rigatelli 1996)

Chevalier and Evariste's younger brother Alfred later made copies of his mathematical papers and may have sent them out, but there is no record of any response.

The melodrama and tragedy of the short life of the headstrong and eccentric Evariste Galois seem largely self-imposed. The circumstances of his death seem to defy rational explanation. One of the letters written on the eve of his death ends: "Preserve my memory, since fate has not given me life enough for the country to know my name" (Rigatelli 1971). Although it took some time before it was properly understood, Galois theory forms one of the most important and beautiful parts of modern algebra.

Ada Byron, Countess of Lovelace

Augusta Ada Byron was born on 10 December 1815. Her father, whom she never met, was the renowned poet George Gordon, Lord Byron. Her mother, Annabella Lady Byron, was considered a mean and selfish hypochondriac by many who knew her. Although she was wealthy, she wasted much of her income on doctors and lawyers. Because she persuaded herself that her husband was plotting to obtain custody of her daughter, she had Ada made a ward in Chancery. Eventually Lady Byron obtained a legal separation from her husband, but she left him soon after Ada was born while he departed for the Continent, where he spent the rest of his life.

Lord Byron, in exile, followed Ada's childhood as closely as he could. He summed up the impression he had received of her temperament in a letter as follows: "Her temper is said to be extremely violent—is it so?—it is not unlikely considering her parentage—my temper is what it is—as you may perhaps divine—and my lady's was a nice little sullen nucleus of concentrated savageness to mould my daughter upon—to say nothing of her two grandmothers—both of whom, to my knowledge, were as pretty specimens of the female spirit as you might wish to see on a summer's day." Later he noted that a description he had received of her disposition and tendencies "very much resembles that of my own at a similar age—except that I was much more impetuous" (Stein 1985). In fact father and daughter had much in common, as we shall see.

Lady Byron tried to keep from her child the scandals involving her father: his incestuous relationship with his half-sister Augusta and his homosexual love affairs. She even concealed his portrait from her. Contrary to what her father had been told, she was a scholarly and quiet child, whose quickness in learning pleased her mother. Ada was educated by Lady Byron (whom Lord Byron had nicknamed "the princess of parallelograms" because of her interest in mathematics) and by tutors closely superintended by her mother. At the age of 8 Ada was amusing herself with building model boats and similar occupations.

Society, courts, and fine gowns all bored Ada, although she was presented at court during the season of 1833. However, she was passionate, quite immoderate, in her intellectual interests: she loved mathematics and music and said, "I am afraid that when a machine, or a lecture, or anything of the kind comes in my way, I have no regard for time, space and any other obstacles." She attended the first of Dr. Dionysius Lardner's lectures on the "difference engine," an early type of calculating machine invented by the eccentric scientist Charles Babbage. Ada understood its workings and saw the great beauty of the invention; her interest in it led to a lifelong friendship with Babbage. She also became friendly with the 44-year-old scientist Mary Somerville, known as the "queen of science," who was impressed by her knowledge of mathematics and astronomy. They often went together to Babbage's house in Marylebone to meet Charles Darwin, Alexander von Humboldt, and other scientific luminaries at his regular Saturday soirées.

The eminent mathematician Augustus de Morgan, who taught Byron calculus by correspondence, described her as having "the potential to become an original mathematical investigator, perhaps of first-rate eminence." However, he warned her mother: "All women who have published mathematics hitherto have shown knowledge, and the power of getting it, but no-one . . . has wrestled with difficulties and shown a man's strength in getting over them. The reason is obvious: the very great tension of mind which they require is beyond the strength of a woman's power of application." (Stein 1985).

In 1835, Ada married William, eighth Lord King, who in 1838 became earl of Lovelace. The scholarly peer was tolerant of his wife's intellectual interests, and they spent most of their time in the country, where they could read and study. Although they had three children, Ada was little involved in their upbringing. "Unfortunately every year adds to my utter want of pleasure in my children," she wrote. "They are to me irksome duties and nothing more" (Stein 1985). Motherhood interested her far less than mathematics, and so her own mother often took care of the children. Babbage was a frequent guest in the Lovelace household; the middle-aged man was flattered by Ada's attention, and she encouraged his chivalrous attachment to her.

Babbage's claim to fame is not based on the difference engine, one of a number of calculating machines that were produced in that period. His great invention was its successor, which he called the "analytical engine." This machine, which was never completed, can properly be described as a computer rather than just a calculating machine, because it could be programmed through the use of punched cards. While the difference engine was designed to work straight through a computational problem, the new machine was designed to make calculations, store the results, analyze what to do next, and then return to complete the project.

In 1842 a treatise on Babbage's invention was published in French. By this time, Ada had developed enough confidence in her mathematical abilities to undertake an English translation, with her own commentary, which expanded the treatise to three times its original length. In a famous and influential metaphor, she wrote that the analytical engine weaves algebraical patterns just as the Jacquard loom weaves flowers and leaves. Although Babbage advised her on substantive matters, Ada was very proprietary about the work, chastising him when he suggested changes. Her husband also helped with the work by copying and making himself useful in other ways. Ada was very proud of her book, which appeared in 1843. She praised her own masterly style and its superiority to that of the original memoir. Her work on the project, especially her program for using it to compute Bernoulli numbers, has earned her the designation of being the first computer programmer. She asked penetrating questions about how the engine might be applied and conjectured that "if it could understand the relations of pitched sounds and the science of harmony the engine might compose elaborate and scientific pieces of music of any degree of complexity and extent" (Stein 1985). She realized the potential of the analytical engine much better than Babbage himself, although it would be over a hundred years before her ideas became a reality.

Ada, who took after her father in many ways, inherited a mercurial temperament that swung precipitously from the ecstatic and grandiose to the melancholic. She also acquired his proclivities for gambling and financial chaos. In 1850 Ada took up betting on horse racing, using a mathematical system she devised with Babbage, and was soon embroiled in legal and financial difficulties as a result. Dangerously in debt, at one point she pawned the Lovelace family jewels. She implored the help of her mother in redeeming them and in concealing the matter from her husband.

The sense of euphoria by which Ada was so easily carried away became a very frequent condition and was succeeded by moods of most intense and crushing fear. These swings from "transcendental elation" to "despair" were often only weeks apart. Like her father, Ada was episodically charged with an awful energy and power and a vastly confident exhilaration of spirit; her grandiosity and occasional delusions rivaled the cosmic sweep of Poe's and Melville's. She wrote that "there is in my nervous system such utter want of all ballast and steadiness that I cannot regard my life or powers as other than precarious." This lack of ballast was reflected also in her grandiose belief that she was "simply the instrument for the divine purpose to act and thro' . . . Like the Prophets of old, I shall speak the voice I am inspired with. I may be Deborah, the Elijah of Science." She had maniacal plans for taking on "the mysteries of the universe, in a way that no purely mortal lips or brains could do." Ada wrote:

> I intended to incorporate within one department of my labours a complete re-duction to a system, of the principles and methods of discovery, elucidating the same with examples. I am already noting down a list of discoveries hitherto made, in order myself to examine into their history, origin and progress. One first and main point, whenever and wherever I introduce the subject, will be to define and classify all that is to be legitimately included under the term discov-ery. Here will be a fine field for my clear, logical and accurate mind, to work its powers upon, and to develop its metaphysical genius, which is not least amongst its qualifications and characteristics. (Stein 1985)

Ada suffered from a series of debilitating illnesses, even allowing for some tendency toward hypochondria. Possibly some of them were psychosomatic. At 3 she caught measles, then a life-threatening disease, and was bedridden for three years. In her seventh year she suffered a mysterious sickness that affected her eyesight and her hearing. At 14 she was afflicted with an illness that deprived her of the use of her legs, but she became normally active afterwards; Stein (1985) suggests that this teenage paralysis might have been hysterical in origin. She also experienced convulsions, like her father. At various times she seems to have suffered from migraine and anorexia, asthma, and gastritis. In 1837, after the birth of her second child, she had a major illness, perhaps cholera. There is always the possibility that some of her illnesses were caused by the treatment she was given, which included the use of opium over an extended period. Opium in the form of laudanum was prescribed for all kinds of medical problems at the time, and to counteract the resulting drowsiness brandy was recommended as a stimulant.

In an appendix to her biography, Dorothy Stein (1985), who is both a psychologist and a computer scientist, discusses Ada's health. As she says, Ada's illnesses, speculations over their causes, and the extent to which they kept her from realizing her early intellectual promise were of such significance to Ada that an attempt to account for these afflictions should be made. Stein discusses various possible explanations for some or all of the problems. One is porphyria, a hereditary disease (the same disorder George III suffered from), which could account for most of them. In the end, however, she concludes that Ada most probably suffered from bipolar disorder, as well as other problems. Since Ada took after her father in her impulsive, aggressive, reckless personality and failure to comply with social norms, it is reasonable to suppose that whatever psychological disorder he had, she had the same. A detailed discussion of his case forms one of the chapters of Jamison (1993a), although an alternative diagnosis is proposed by one of the present authors (Fitzgerald 2001). Ada died of natural causes, but as she approached the age of 36, she was well aware that her father had died at precisely that age, as had his father before him. She had suffered from cervical cancer since 1850; when she died in 1852, she too was 36.

Bernhard Riemann

Georg Friedrich Bernhard Riemann was born in a remote part of the Kingdom of Hanover on 17 September 1826. His childhood was spent in the obscure village of Quickborn, where his father was pastor, and until he was almost 30 he regarded it as home. By the age of 6 he was solving arithmetical problems under the tutelage of his father. At 14 he entered the senior class of the gymnasium at Hanover, where he was deeply unhappy; two years later he transferred to the one at Luneburg, much nearer home, where he continued until he was 19. The director of the school allowed him to study in his personal library, which included some scientific classics.

In 1846, in accordance with his father's wishes, Riemann matriculated at the Georgia Augusta in the faculty of theology. His interest in mathematics was by then so strong, however, that he persuaded his father to allow him to transfer to the appropriate faculty, that of philosophy. After his first year in Göttingen, Riemann moved to Berlin, where Lejeune Dirichlet, Gotthold Eisenstein, Carl Jacobi, and Jakob Steiner were professors. While Eisenstein was particularly friendly to Riemann, it was Dirichlet whose lectures had the greatest influence.

Two years later, Riemann returned to Göttingen and spent eighteen months working on mathematical physics, especially electrodynamics. After two more years of intense mathematical work, Riemann habilitated at the end of 1853. Part of the examination was a trial lecture, in his case on the foundations of geometry. Gauss was one of the examiners—perhaps the only one who could appreciate the depths of Riemann's thought. It exceeded all Gauss's expectations and greatly surprised him; afterward he spoke of it with the highest praise and rare enthusiasm. Riemann wrote home:

> I became so absorbed in my investigation of the unity of all physical laws that when the subject of the trial lecture was given to me I could not tear myself away from my research. Then partly as a result of brooding on it, partly from staying indoors too much in this vile weather I fell ill; my old trouble recurred with great pertinacity and I could not get on with my work. Only seven weeks later when the weather improved and I got more visual stimulation I began feeling better. For the summer I have rented a house with a garden, and since doing so my health has not bothered me. Having finished two weeks after Easter a piece of work I could not get out of, I began at once working on my trial lecture and finished it around Pentecost, I had some difficulty getting a date for my lecture right away and almost had to return to Quickborn without having reached my goal. For Gauss is seriously ill and the physicians fear that his death is imminent. Being too weak to examine me, he asked me to wait until August, hoping that he might improve, especially as I could not lecture anyhow until the autumn. Then he decided anyway on the Friday after Pentecost to set the lecture for the next day at eleven thirty. On Saturday I was happily through with everything.

So Riemann finally obtained the right to teach publicly. He gradually overcame his initial shyness and established some rapport with his audience. He wrote to his father, "I have been able to hold my classes regularly. My first diffidence and constraint have subsided more and more, and I get accustomed to think more of the auditors than myself and to read in their expressions whether I should go on or explain the matter further." However, according to Dedekind, Riemann's lecturing style did not improve. All his life he found it difficult to relate to other people.

Riemann was deeply attached to his family and maintained the closest contact with its members throughout his life. Timid and reserved by nature, he felt at ease only in their company. In 1855 his father died, and soon afterward he lost one of his sisters. The other three went to live with his brother in Bremen, so Riemann no longer had a home to retreat to in Quickborn. These events, in addition to nervous exhaustion brought on by overwork, led him into depression. He went off to the nearby Harz mountains, pursued by Dedekind, who wrote to his sister:

> one must do all one can to save so excellent and scientifically most important person as Riemann from his now utterly unhappy state, but he must not notice this intention too clearly; it has always been difficult to do him a favour. And the only way one could get him to accept a favour was to persuade him that one did it as much for one's own sake as for his; he hates to trouble other people. He has done the strangest things here only because he believes that nobody can bear him.

There were further misfortunes on the way. First Riemann's brother died, and as a result the three sisters who had been living in Cologne came to Göttingen, where providing for them imposed a heavy additional burden on his limited finances. Before long one of the sisters also died. However, Riemann's career was progressing well, as the importance of his research gained recognition. When Dirichlet died, Riemann, who was already 36, was appointed his successor; as an *ordinarius* he could afford to marry and start a family. His bride, Elise Koch, was a friend of his sisters. The wedding took place in 1862, and a daughter was born the following year. Soon, however, he developed pleurisy, the first indication that tuberculosis had taken hold, and before long Dedekind would write that "he could only lecture with great difficulty. One could see his tiredness and lassitude, his thoughts frequently failed him, and he was unable to explain the simplest things."

Much of the last years of Riemann's life was spent with his family almost entirely in Italy, primarily for health reasons but also so that he could pursue his interests in art and antiquities. He already knew some of the Italian mathematicians, because several of them had visited Göttingen.

When they offered him a position at Pisa he declined it for reasons of health. Another of his sisters died, leaving him with only one sibling. He gave up attempting to lecture and returned to Italy late in June 1866 with his family. They spent a few weeks at the beautiful northern end of Lake Maggiore, but Riemann's strength was rapidly declining and he felt the end was near. Dedekind, in his obituary, describes how, on the day before his death, resting under a fig tree, Riemann's soul filled with joy at the glorious landscape; he was working on some ideas that he did not live to complete. Riemann died on 20 July 1866, less than 40 years old. After his death, a note was found among his papers stating the famous Riemann hypothesis about the zeroes of the zeta function, which remains an open question to this day.

In boyhood Riemann's health had been poor and, in general, sickness and early death haunted all the members of the close-knit family to which he belonged. It seems almost certain that tuberculosis was the cause of his death. Until the discovery of the bacillus responsible, the disease was thought in northern Europe to be hereditary, although further south it was known to be transmitted by infection. In just fifteen years of creative activity he made enormous contributions to mathematics. However, the value of his work lies not only in these achievements but in his pioneering work on topology, which was to stimulate mathematicians for years to come. In the ranks of outstanding mathematicians of the nineteenth century, Riemann occupies a special place.

Chapter 6 } Cantor, Kovalevskaya, Poincaré, and Hilbert

Georg Cantor

Georg Woldemar Cantor, the father of the mathematician, was a native of Copenhagen who in his youth moved to St. Petersburg. There he became a successful stockbroker and married Maria Anna Bohm Meyer, who came from a musical family. Their son Georg Ferdinand Ludwig Philipp was born on 3 March 1845, the first of four children. He was raised in an intensely religious atmosphere; his mother was a Roman Catholic but his father was a staunch Lutheran. On the occasion of his son's confirmation, his father, a sensitive and gifted man, wrote a letter to him that Cantor would never forget. It foreshadowed much that was to happen to him in later years:

> No-one knows beforehand into what unbelievably difficult conditions and oc-cupational circumstances he will fall by chance, against what unforeseen and unforeseeable calamities and difficulties he will have to fight in the various sit-uations of life. How often the most promising individuals are defeated after a tenuous, weak resistance in their first serious struggle following their entry into practical affairs. Their courage broken, they atrophy completely thereafter, and even in the best case they will still be nothing more than a ruined genius. But they lacked the steady heart, upon which everything depends! Now, my dear son! Believe me, your sincerest, truest and most experienced friend—this sure heart, which must live in us, is a truly religious spirit.
>
> But in order to avoid as well all those other hardships and difficulties which inevitably rise against us through the jealousy of and slander by open or secret enemies in our eager aspiration for success in the activity of our own speciality or business; in order to combat these with success one needs above all to ac-

quire and to appropriate the greatest amount possible of the most basic, diverse technical knowledge and skills. Nowadays these are an absolute necessity if the industrious and ambitious man does not want to see himself pushed aside by his enemies and forced to stand in the second or third rank.

In 1856, when young Cantor was 11, the family moved from Russia to Germany. They lived in Wiesbaden to start with, and from the age of 15 he attended the local Gymnasium. He showed all-around ability, but mathematics and science were his strongest subjects, and his father decided he should train as an engineer. With this end in view, Cantor began his higher education at the Polytechnikum in Zurich. Already there were signs of the mental instability that was to worsen as he grew older. In Zurich, the young man worked long into the night, endangering his health. Lack of sleep left him tired and eventually so despondent that his father became worried about his melancholic condition.

Next, Cantor's father, who was dying of tuberculosis, agreed to let him transfer to the University of Berlin and study mathematics instead of engineering, where he proved to be a good but not exceptional student. After obtaining his doctorate in 1867, at the age of 22, Cantor became a school-teacher for a short while and then *privatdozent* at the University of Halle, not far from Leipzig. Fortunately his wealthy father had left him well off, since it was five years before he became associate professor and another seven before he was promoted to full professor. Although the stipend was not generous, he remained at Halle for his entire career.

Those who knew him in his prime described Cantor as energetic, forceful, and volatile. One in particular described him as a man of loud and intense character, true to friends, helpful when needed, noble and generous. Another described him as a man of imposing stature, witty, spirited, amiable, and in his youth particularly lively and stimulating in conversation. Yet another described him as one of the most stimulating of Germany's mathematical personalities; his presence at any congress or meeting was always an enticing attraction; his mind was as imaginative and sparkling as it was temperamental and explosive. On one occasion Karl Weierstrass was entertaining a number of mathematical friends, including Cantor, when apparently without warning, Cantor exploded with rage over the fact that he had not been offered the position given to Felix Klein at Göttingen three years earlier. Such a swift and violent outburst foreshadowed the paranoid behavior of his later years.

Cantor described himself as "rather artistically inclined," and at times he appeared regretful that his father had not let him become a violinist. He married Vally Guttman, a friend of his sister's, in 1874: they had six children. During their honeymoon in the Harz mountains the couple happened to meet Richard Dedekind, who became a close friend and mathematical confidant of Cantor's.

After settling down at Halle, Cantor started to think deeply about the difference between denumerably infinite sets like the rational numbers and continuous sets like the real numbers. He developed his increasingly revolutionary ideas in a series of papers on what we now call transfinite set theory, published between 1879 and 1884. Nowadays the logical position is much better understood, but students of mathematics still feel uneasy when they meet, for the first time, the ideas that were so controversial when Cantor introduced them. When they appeared, Cantor's theories were generally received with skepticism, often hostility, particularly from Leopold Kronecker, who tried to prevent or at least delay their publication. Cantor would have liked to return to the more mathematically stimulating environment of Berlin, but the influential Kronecker ensured that he did not. Other influential mathematicians were also unsympathetic, but there were exceptions. David Hilbert described Cantor's theory of transfinite numbers as one of the greatest achievements of the human spirit.

For ten years, Cantor, discouraged by the hostile reception of his theories, turned away from mathematics. He published in philosophical journals, and his correspondence reveals a preoccupation with matters such as Rosicrucianism, freemasonry, and theosophy, and also with literary questions such as the Bacon-Shakespeare controversy. When he returned to mathematics it was with another major work, the synoptic *Beiträge*, which stimulated heated polemics between widely separated camps of mathematical opinion. Partly to try to secure a fairer hearing for his controversial ideas, Cantor was instrumental in the foundation of the German Mathematical Society in 1889. By that time other mathematicians were taking up his ideas and, in the hands of Gottlob Frege, and later Kurt Gödel, the implications of the deep questions he raised became better understood.

The second half of Cantor's life was marred by recurrent episodes of mental illness, as the cyclothymic episodes of his youth turned into classic bipolar disorder. According to the medical records that survive, the attacks all began suddenly, usually in the autumn, and exhibited phases of excitement and exhilaration. They ended suddenly in the following spring or early summer and were sometimes followed by what we now understand to be the depressive phase, in which he would sit silent and motionless for hours on end.

Cantor's first major mental breakdown had occurred in the spring of 1884, just after he returned from a successful trip to Paris. His eldest daughter Elise, only 9 at the time, was bewildered by the incomprehensible change in her father's personality, the swiftness with which his entire manner had been transformed. As soon as he recovered he made an attempt, quite cordially received, at a reconciliation with Kronecker. He devoted most of his time to literary-historical problems, hoping to prove that Francis Bacon was the true author of Shakespeare's plays. He also intensified his study of the scriptures and the fathers of the church. He suddenly desired to teach philosophy at Halle rather than mathematics. In an amusing letter, Sonya Kovalevskaya, whose profile appears in this chapter, describes what happened when Cantor tried giving a course on the philosophy of Leibniz:

> In the beginning he had 25 students, but then little by little, melted together first to 4, then to 3, then to 2, finally to a single one. Cantor held out nevertheless and continued to lecture. But alas! One fine day came the last of the Mohicans, somewhat troubled, and thanked the professor very much but explained he had so many other things to do that he could not manage to follow the professor's lectures. Then Cantor, to his wife's unspeakable joy, gave a solemn promise never to lecture on philosophy again!

Eventually the breakdowns began anew, grew more intense, lasted longer, and occurred with greater frequency. In 1899, the second time there are records of his hospitalization for mental instability, Cantor was on sick leave for the autumn term. He wrote to the Ministry of Culture saying he was anxious to relinquish his professorship at Halle for a more modest position. He emphasized his qualifications, his knowledge of history and literature, his publications on the Bacon-Shakespeare question, and even added that he had come upon certain information concerning the first king of England "which will not fail to terrify the English government as soon as the matter is published." He asked for a reply within two days, writing that otherwise, as a born Russian, he could apply for a position in the tsar's diplomatic corps. Nothing came of this; he remained at Halle as before.

For part of 1899 Cantor was again hospitalized, and before the year was out his youngest son Rudolf died suddenly, four days before his thirteenth birthday. The boy had been frail in childhood, but then began to grow stronger. He was gifted musically—his father hoped he might follow in the family tradition and become famous as a violinist. Disappointed and unhappy over his own choice of profession, Cantor wanted something better for his son.

Cantor managed to maintain his mental equilibrium for the next three years, until he was again hospitalized and relieved of his teaching duties for the winter term of 1902–1903. At the third International Congress of Mathematicians in Heidelberg, his theories came under attack. The strain was too much for him; he was soon back in the hospital on sick leave. Thereafter, he spent increasingly long periods at the Nervenklinik in Halle, being admitted for the last time in May 1917. He was reluctant to go there and kept asking his family to take him home. On 6 January 1918, Cantor died, apparently of heart failure, at the age of 72. After his death Hilbert said, "Cantor has created a paradise from which nobody shall expel us."

Various aspects of Cantor's career have been misrepresented in the literature, and his recent biographers have found it difficult to rectify matters. At certain points, particularly in 1904, there seemed to be links between periods of emotional upset and impasses in Cantor's mathematics; in his lifetime these links were greatly exaggerated. There is good evidence for Cantor's mental illness being endogenous and having nothing to do with the opposition to his mathematics, intense though that was. Although it is claimed that Cantor's theory was upset by a number of "entirely unexpected" contradictions, Cantor was well aware of these "paradoxes" from the outset, although perhaps he did not appreciate their full implications until later.

At least in Cantor's own mind, his periods of depression and seclusion served a unique and generative purpose, providing rest and quiet during which he could make great progress in many facets of study by mere reflection. For example, after his release from the hospital in 1905, he had "an inspiration from above, which suggested to me a renewed study of our bible with opened eyes and with banishment of all previous misconceptions." As a result Cantor produced his pamphlet *Ex oriente lux*, in which he argued that Jesus Christ was the natural son of Joseph of Arimathea. During the long periods of seclusion his mind was left free to ponder many things, and in the silence he could perceive the workings of a divine muse and could hear a secret voice from above that brought him both inspiration and enlightenment. Apparently Cantor's inner voice knew more than the details of Christian history; his muse was also a mathematician.

In the words of Joseph Dauben:

> There can be no mistake about Cantor's identification of his mathematics with some greater absolute unity in God. Even before his first breakdown of 1884, Cantor had told Mittag-Leffler that his transfinite numbers had been communicated to him by a more powerful energy; that he was only the means by which set theory might be made known. Cantor believed that God endowed the transfinite numbers with a reality that made them very special. Later generations might forget the philosophy, smile at the abundant references to St. Thomas and the fathers of the Church, overlook his metaphysical pronouncements and miss entirely the deep religious roots of Cantor's later faith in the veracity of his work. His forbearance, as much as anything else he might have contributed, ensured that set theory would survive the early years of doubt and denunciation to flourish eventually as a vigorous, revolutionary force in scientific thought of the twentieth century.

Sonya Kovalevskaya

The outstanding female mathematician of the nineteenth century was the Russian Sonya Kovalevskaya, the protégée of Karl Weierstrass and Gøsta Mittag-Leffler, whose unconventional life came to an untimely end just when she seemed to have surmounted all of the obstacles she faced in her career. Although she died relatively young, she had become accepted, through her personality as much as her ability, as a member of the research community in a way that no woman had been before. When she died there was great curiosity about her brain, which was found to be unusually large, and this was claimed to be the reason for her exceptional mental capacity. She was the first woman in the world to get a doctoral degree in mathematics in the modern sense of the term, and she was among the first women to earn advanced degrees in any discipline. She was not only a mathematician but also had considerable literary talent. We know more about her personality than we do about most of her contemporaries, partly through her writings, which were semi-autobiographical, and partly through the reminiscences of others, notably her friends Julia Lermontova and Anne-Charlotte Leffler. For a detailed account of her eventful life, the excellent biography by Don Kennedy (1983) can be recommended, for an outline, the biographical sketch by Ann Koblitz (1987). The main source of this profile is Kennedy's biography, from which the quotations are taken.

The future mathematician was born Sofia Vasilievna Krukovskaya in Moscow, on 15 January 1850 by the Gregorian calendar. Her father, a landowner, was a general officer in the Russian artillery. Her mother came from a family of German scholars who had settled in Russia in the time of Catherine the Great. Sofia had a brother and an elder sister Anna, known in the family as Aniuta, just as Sofia was always called Sonya. As a child Sonya was rather shy and self-contained, but could be quite stubborn and capable of rages. At times she was lively and energetic, seldom in repose, but she lacked the suppleness and grace of her sister. She was careless in her dress and had a limp handshake. She sometimes created a feeling of unease by fixing her penetrating gaze on a visitor as though analyzing or passing judgment on the individual. Like most girls of her class, as a child she was left mainly in the care of nurses and governesses. A highly intelligent child, she was largely self-taught. When she was old enough, her well-educated father arranged for her to have some tuition in science and mathematics; in the latter she progressed to a level achieved by few men at that time.

In the 1860s, the nihilist philosophy became current in Russian educated circles. According to this theory, the traditional structure of tsarist society had to be changed. The nihilists believed in the power of education to improve civic society and felt that the most useful branches of knowledge were the natural sciences. They thought that science would hasten the peaceful social revolution that they all considered inevitable, and they viewed a scientific career as an active blow against backwardness and autocracy. Moreover, the nihilists coupled this faith in science and social revolution to a firm belief in the equality of women. Universities in Russia, like those of all Europe at the time, were closed to women, although Russian women, fired by nihilistic philosophy, expected the universities to open to them before long. When this optimism proved unfounded, they considered studying abroad. However, by law, women were not allowed to leave Russia except in the company of a husband or chaperone. The nihilist solution to this problem was what they called a fictitious marriage—a marriage in law that was never intended to become a real union.

Sonya had been brought up with the expectation that her future would be settled by marriage with a young man of suitable wealth and family position. However, led by her older sister Aniuta, whom she adored, Sonya encountered the nihilist philosophy at an early age. The sisters found its emphasis on natural sciences, revolutionary social change, and equality of opportunity for women congenial. Sonya remained true to nihilist precepts throughout her life, and always identified herself as a member of the movement. At the age of 18, much against her father's wishes, she contracted a fictitious marriage with a fellow nihilist, Vladimir Kovalevsky, whose profession was paleontology; her surname then became Kovalevskaya.

After her marriage, Kovalevskaya made a vain attempt to enter an institution of higher education in St. Petersburg, before deciding to leave the country with her sister and husband. The following year the sisters left Russia with Vladimir and reached Vienna. There they separated, with Aniuta going on to Paris to join some political activists, while Sonya and Vladimir went to Heidelberg. Not without difficulty, Kovalevskaya obtained permission to attend courses at the university. She studied science with Robert Bunsen, Hermann von Helmholtz, and Gustav Kirchhoff, and mathematics with Emil Du Bois-Reymond and Leopold Königsberger, former students of Weierstrass. This was a strenuous program, but even so the newlyweds found time to travel to England, where Vladimir met the biologists Charles Darwin and Thomas Huxley, while Kovalevskaya visited George Eliot, with whom she found she had much in common, and engaged in discussion with the philosopher Herbert Spencer about the capacity of women for abstract thought.

Before long, the fictitious marriage ran into various kinds of difficulties, and the couple separated. In the autumn of 1870, Kovalevskaya went to Berlin to establish contact with Weierstrass himself. When the great analyst tested her with a few problems, he was so impressed by her answers that he began giving her private lessons, covering much the same ground as he did in his university courses, which Berlin regulations prevented women from attending. He called her "the best pupil I have ever had." She persuaded him to promise that he would not accept another woman for private study. As she was to say later, "These studies had the deepest possible influence on my mathematical career. They determined finally and irrevocably the direction I was to follow in my later scientific work; all my work has been done precisely in the spirit of Weierstrass" (Kennedy 1989).

Weierstrass's role in Kovalevskaya's scientific and personal affairs went far beyond the usual student-teacher relationship. He found her captivating: a dangerous woman. In her he found a "refreshingly enthusiastic participant in all his thoughts, and much of his thinking became clear in his conversations with her." It seemed to him that "we had been close throughout my entire life and never have I found anyone who could bring me such understanding of the highest aims of science and such joyful accord with my intentions and basic principles as you." While on holiday in 1873 he wrote to her:

> During my stay here I have thought about you very often and imagined how it would be if only I could spend a few weeks with you, my dearest friend, in such a magnificent natural setting. How wonderful it would be for us here—you with your imaginative mind, I stimulated and refreshed by your enthusiasm— dreams and flights of fancy about so many puzzles that remain for us to solve about finite and infinite spaces, the stability of the solar system, and all the other great problems of the mathematics and physics of the future. However, I learned long ago to resign myself if not every beautiful dream comes true.
> (Kennedy 1983)

Meanwhile, Sonya's sister Aniuta had become much involved with the Paris radicals, including one named Victor with whom she was living. They got into serious trouble with the authorities, so much so that Victor was at risk of summary execution. With great difficulty, Sonya and Vladimir, who had reconciled, managed to enter Paris, then under siege, in the hope of helping Aniuta and Victor to escape. For several months they experienced the life of the Paris Commune. Sonya's parents had also gone to Paris out of concern for Aniuta. Her father used his influence with the military to enable Victor to flee to Switzerland, while Aniuta escaped to England before joining Victor in Switzerland, where they were married.

After these dramatic events, Kovalevskaya returned to Berlin and resumed her mathematical studies with Weierstrass. He gave her every encouragement and, judging by the letters he sent her, there was affection as well. To a large extent, Russian nihilist women of Sofia's generation opened higher educational institutions in continental Europe to women. They were among the first officially enrolled students in Zurich, Berne, Geneva, Heidelberg, and elsewhere. Weierstrass encouraged Kovalevskaya to aim for a doctorate: although this would have not been possible at Berlin, the regulations at the Georgia Augusta did not exclude women and, in the autumn of 1874, she was awarded the degree *summa cum laude* for the work she had done in Berlin.

Kovalevskaya was almost 25. She and her husband had reconciled during 1873 and made plans for a future together. He too had achieved a doctorate. They returned to Russia in the autumn of 1874 and were feted by the intelligentsia of St. Petersburg, not so much the scientists as the artists, scholars, writers, and idealists. Never very far from radical circles, they devoted much of their time to participating in various projects for reform, usually fruitlessly. Kovalevskaya wrote an acclaimed memoir, *Memories of Childhood,* and the novella *A Nihilist Girl,* apparently based on her early life. She also wrote essays, one-act plays, and other literary works.

However, for Sonya and Vladimir disillusionment with life in their homeland was not slow in coming. To teach above the elementary level in Russia, even with a German doctorate, it was necessary to pass the magisterial examination, which women were not allowed to take. Vladimir failed at his first attempt, but finally passed in 1875. Even then he failed to secure a post that he considered appropriate. Vladimir had an extremely restless nature and was constantly being carried away by new ideas and plans. He hoped to make his fortune through business speculations of doubtful legality.

Vladimir believed that Sonya "absolutely cannot be a mother; this would simply ruin her, and she herself fears this; it would tear her away from her work, making her unhappy, and besides I think she would make a bad mother; there isn't a single maternal instinct in her, and she simply hates children" (Kennedy 1983). Nevertheless, Sonya and Vladimir finally consummated their union after seven years of platonic relationship, and in 1878 she gave birth to a daughter, Sophia Vladimirovna, nicknamed "Fufa."

Shortly thereafter, Kovalevskaya met Mittag-Leffler, a fellow disciple of Weierstrass, who was stunned by her spirituality, charm, and mathematical knowledge. He wrote:

> More than anything else in St. Petersburg what I found most interesting was getting to know Kovalevskaya. As a woman she is fascinating. She is beautiful and when she speaks her face lights up with such an expression of feminine kindness and highest intelligence that it is simply dazzling. Her manner is simple and natural, without the slightest trace of pedantry or pretension. She is in all respects a complete "woman of the world." As a scholar she is characterized by her unusual clarity and precision of expression . . . I fully understand why Weierstrass considers her the most gifted of his students. (Kennedy 1983)

Mittag-Leffler was so impressed by Kovalevskaya's ability that he attempted to find a position for her at the University of Helsinki, where he was professor. He soon discovered that her Russian nationality, combined with her reputation as an active radical, made this unlikely.

After several years, Kovalevskaya had started to resume her mathematical activities, feeling that the effort she had put into her education must not be wasted. Leaving her husband in Russia, she returned to Berlin in 1881 with her daughter. There Weierstrass suggested a new mathematical project for her to work on, while Mittag-Leffler continued his efforts to obtain a position for her. The next year she moved to Paris, leaving Fufa in the care of relatives in Odessa. Meanwhile her husband was desperately trying to sort out his financial affairs. Depressed by his inability to start a career and a realization that his business partners were swindlers who were landing him deeply in debt, he lost all interest in work and people and withdrew from society. Finally, in April 1883, he took his own life. After Kovalevskaya recovered from the initial shock of this event, she returned to Russia, where she managed to clear her late husband's name, put his affairs in order, and make temporary arrangements for the care of their daughter.

About this time universities in several countries were preparing to admit women to their faculties; the first to do so was in Sweden. Mittag-Leffler had moved to the new Høgskola in Stockholm and managed to arrange for her to join him there, where she worked as a *privatdozent*. Although she was new to the lecture hall and her German was less than fluent, she quickly overcame her initial lack of confidence. At the end of the course, she was toasted by the students and presented with a framed photograph of herself. Her probationary term was such a success that Mittag-Leffler was able to persuade the Høgskola to appoint her to a five-year professorship.

Kovalevskaya also became one of the early editors of the *Acta Mathematica*. Stockholm was then something of a mathematical backwater, but through her work on the *Acta,* supplemented by occasional visits to Berlin and Paris, she was able to keep in contact with many of the leading mathematicians of the period. At first she was happy enough in the Swedish capital. Her work was going well. Socially she was a sensation, the star of an endless series of dinner parties. Although the fact that she had left her daughter behind in Russia caused some adverse comment, as a widow she was acceptable in staid Swedish society in a way that she would not have been if her husband was still alive but separated from her. Mittag-Leffler's sister Anne-Charlotte had become a close friend. They wrote a play together, *The Struggle for Happiness,* which was favorably received when it was performed in Moscow. Kovalevskaya hoped that the St. Petersburg Academy, of which she was already a corresponding member, might elect her to full membership, but there was too much opposition because she was a woman.

Sonya's sister Aniuta had been ill for some time, and in the summer of 1886 her condition deteriorated. She and her husband were expelled from Russia and returned to Paris, where she died following an unsuccessful surgical operation. By this time Kovalevskaya was nearing the end of her five-year appointment at the Høgskola. Fortunately, the theme of the prestigious Bordin prize competition of 1888, set by the Paris Academy, was on the motion of a rigid body, and her entry, on the dynamical theory of what became known as the Kovalevskaya top, brought her international fame. As a result the Høgskola made her a full professor, with tenure, the following year, and her future seemed assured.

However, the events of the previous few years, especially the death of Aniuta, had taken their toll. Leaving Fufa behind with the Mittag-Lefflers, Kovalevskaya spent the first part of 1890 in Paris trying to recuperate. She returned to Stockholm for the autumn term and then went on holiday to Genoa. On the return journey she became ill, possibly with influenza. Kovalevskaya died on 10 February 1891, in her forty-second year, of a pulmonary infection. Her untimely death caused Weierstrass much grief. He burned all of her letters to him: only the draft of one of them has survived. It deals with her arrival in Stockholm and first reactions to the place where she spent the last years of her life.

According to Kennedy, all those who knew her remembered her as a woman of great spirit and originality. Along with a very masculine energy and often tough inflexibility, she showed a very feminine personality in many things, and wherever she went she attracted men. She was very poor at managing money. Her power of understanding and sympathizing with the thoughts of others, noted by her friend Anne-Charlotte, seems to rule out any suggestion of Asperger syndrome, although her personality displayed some Asperger traits, and perhaps these helped her to achieve success in the face of many difficulties. In adult life she kept swinging from brief periods of delight to longer periods of depression, suggesting cyclothymia.

After Kovalevskaya's death, Anne-Charlotte prepared a volume of reminiscences, which was published in Swedish. For this she obtained some very interesting material from Kovalevskaya's close friend Julia Lermontova, who recalled that Sonya's love for mathematics and her unusually sympathetic personality attracted everyone she met. Lermontova's remembrances have been translated into English in the Kennedy biography, including the following passage:

> There was something in her absolutely adorable. Her moral image was completed by a deep and complicated spiritual psyche such as I never encountered again in anybody. She couldn't tolerate failure. No sooner did she think of a goal than she would strive with all her strength to achieve it, using every means at her command. For this reason she always achieved what she wanted, except on the occasions when emotion interfered, at which times, in some strange fashion, she lost the acuteness of judgement habitual to her. She demanded too much always from the one who loved her, and whom she loved in turn, always wanting to take by force what the person loving her would himself have given willingly if she had not used such passionate insistence. She always felt an irresistible need for gentle intimacy, a need for having a person always with her to share everything with her, and at the same time she made life impossible for the person in such a close relationship. She was too restless in character, too lacking in harmony, to find satisfaction for very long in a quiet life full of affection of which she apparently dreaded. Besides this she was too individual in character to pay sufficient attention to the yearnings and inclinations of the person living with her. She once said that like a chameleon she tended to take on the colouration of the person she was with. In concentrating on one person at a time she expected similar concentration by that person upon herself, and when the other person's attention wandered off to other persons or other interests she felt rejected and unhappy.
>
> Her ability over many hours to devote herself to concentrated mental labour without leaving her desk was really astounding. And when, having spent the entire day impressing work, she finally pushed away her papers and arose from her chair, she was always so submerged in her thoughts that she would walk back and forth with quick steps across the room, and finally break into a run talking loudly to herself and sometimes breaking into laughter. At such times she seemed completely separated from reality. Carried by fantasy beyond the borders of the present, she would never consent to tell me what she was thinking on these occasions. She slept very little at night and had disturbing dreams. Often she would awake suddenly from some fantastic dream and would ask me

to sit with her. She readily related her dreams, which were always very original and interesting. Not infrequently they were like visions to which she ascribed prophetic significance and that often actually came true. In general she was distinguished by an extremely nervous temperament. She was never at peace, always setting difficult goals for herself, always wanting passionately to attain them. Despite this, I never saw her in so depressed a state of mind as when she had achieved a particular goal. It seemed the reality of the achievement never corresponded to what she had imagined. While working she caused little pleasure in those around her, being wholly immersed in her work. Yet when one observed her, melancholy and sorrowful in her complete success, one involuntarily felt a deep sympathy for her. These constant changes in disposition, from sadness to happiness, made her most interesting to know.

Henri Poincaré

Henri Poincaré was born in the French city of Nancy on 29 April 1854. Both of his parents came from bourgeois families that had lived in Lorraine for generations. His forebears included a number of distinguished people. Poincaré's father Léon was a physician and professor of medicine at the University of Nancy. Henri and his younger sister Aline were at first educated at home by their mother Eugenie (née Lannois), who "watched over them with a solicitude full of intelligence." Henri learned to talk very early but had trouble at first expressing the rush of his thoughts. A severe attack of diphtheria at the age of 5 left him unable to speak for almost a year. He enjoyed playing with other children but disliked the more boisterous games. When on holiday in the country he wanted to see everything, understand everything, and explain everything. He was particularly fond of animals. Poincaré tended to walk rapidly and enjoyed dancing. However, his motor coordination was poor and he was remarkably maladroit; it was said he could write or draw just as badly with his left hand as with his right. He suffered from allergies and insomnia and was severely myopic: he recognized other people by the sound of their voices rather than by sight. Also, he could be extremely obstinate.

Poincaré grew up in a comfortable intellectual environment. With ample time to read and study, he made rapid progress. At the age of 8, when he began formal schooling, his unusual ability showed first in languages, both ancient and modern, but by the end of his school career his awesome mathematical talent was already apparent. His schooling was interrupted by the Franco-Prussian War, during which his home province of Alsace-Lorraine bore the brunt of the German invasion. He accompanied his father on ambulance rounds at this time, becoming a lifelong French patriot as a result.

At school Poincaré carried off first prize in the Concours Général for elementary mathematics and came first after a brilliant performance in the entrance examination for the École Polytechnique in 1873. He also took the examination for the École Normale Supérieure and was placed fifth on the list of candidates. More than the Polytechnique, Normale was the cradle of the French intellectual elite, a place where the country's brightest and most gifted youth could spend three highly formative years studying, debating, and socializing in a close-knit community of scholars. However, this may not have appealed to Poincaré, who was planning to make a career in engineering, and he chose the more scientifically oriented Polytechnique.

At the Polytechnique, Poincaré again made rapid progress, although clumsiness in drawing and experimental work cost him first place in the final examination. He went on to the École des Mines for the next three years and, after qualifying, worked as a mining engineer in Vésoul. However, this did not last long: his career was to be in mathematics rather than engineering. He wrote a doctoral thesis on the properties of functions defined by differential equations, later published in the *Journal de l'école polytechnique,* which secured him a position at the University of Caen in Normandy. There Poincaré made his first important discoveries, on theta-fuchsian functions. For the next few years he worked feverishly to develop his ideas in competition with the young German mathematician Felix Klein. There were some reservations about his style—undisciplined and lacking in rigor, though very readable—but his brilliance was not contested. In 1881 Poincaré returned to the French capital, at the age of 27, to be a lecturer in mathematical analysis at the Sorbonne. Although he was a conscientious teacher, his lectures were described as "infinitely austere." It must have been around this time that the British mathematician J. J. Sylvester visited Poincaré at his "airy perch" in the rue Gay-Lussac and was astonished when, after having toiled up three flights of narrow stairs leading to the study, he found that a mere boy, "so blond, so young," was the author of the deluge of papers that had heralded him as a successor to Cauchy.

Later Sylvester got to know Poincaré quite well. On one occasion, several people, including Poincaré, had been invited to a dinner in honor of Sylvester. When Poincaré arrived, Sylvester monopolized his attention without giving him time even to greet his hostess: "I have a beautiful theorem to show you," said Sylvester, and proceeded to demonstrate it. From that moment Poincaré said not a word; he ate his meal like an automaton. After dinner Poincaré recovered his awareness of the outside world and descended on Sylvester, exclaiming, "but your theorem is false!" and proving it to him on the spot. On other occasions, however, Poincaré was described as "affable," even "charming." He found time for an active social life, and once he settled down in Paris he married Jeanne Louise Marie Poullain d'Andecy, who had a similar social background to his own; they would have four children, a son and three daughters.

Within five years of his return to Paris, Poincaré was appointed professor of mathematical physics and calculus of probability at the Sorbonne, then, ten years later, professor of mathematical astronomy and celestial mechanics —the chair he retained, winning higher and higher honors, until the end of his life. He also gave courses on astronomy at the École Polytechnique and on electricity at the École Professionelle Supérieure des Postes et Telegraphes. In 1887, at 32, he was elected to the Paris Academy as a result of his work on automorphic functions, and six years later to the Bureau des Longitudes. The following year he was elected to the Royal Society of London, the first of many such honors; he later became the first holder of the society's Sylvester medal. Like Leonhard Euler, he wrote fluently and copiously on every part of mathematics; in fact he surpassed even Euler in this. He also wrote several books on science and philosophy that were bestsellers in the early part of the twentieth century and are still read today. In 1908, he was elected to the Académie Française, the literary section of the Institut de France.

A picture of the great man at work can be found in a letter written by his nephew, the philosopher Pierre Boutroux:

> Poincaré's normal practice was to work on mathematics for no more than four hours a day; 10 am to noon and 5–7 pm. Evenings were reserved for journal reading. In his peaceful study in Paris or in the shade of his garden in the Lozère, Henri Poincaré would sit for hours every day in front of a pad of ruled paper, and one saw the sheets being covered, with a surprising regularity, in his delicate and angular hand-writing. Almost never an erasure, very rarely a hesitation. After some days a lengthy memoir will be finished, ready for the printer,

and my uncle from then on was only interested in it as something in the past. He could scarcely be persuaded to cast a quick glance at the proof-sheets when they were sent to him by editors. (Darboux 1916/1956)

In research Poincaré placed far more reliance on intuition than is typical; apparently that was why he did not believe in collaborating with other mathematicians. Boutroux added:

> It was often observed that Henri Poincaré kept his thoughts to himself. Unlike certain other scientists, he did not believe that oral communication, the verbal exchange of ideas, could favour discovery . . . my uncle regarded mathematical discovery as an idea which entirely excluded the possibility of collaboration. The intuition, by which discoveries are made, is a direct communion, without possible intermediaries, with the spirit and the truth . . . He thought in the street as he went to the Sorbonne, while he was attending some scientific meeting or while he was taking one of his habitual grand walks after lunch. He thought in his antechamber or in the hall of meetings at the institute, while he walked with little steps, his physiognomy tense, shaking a bunch of keys. He thought at the dinner table at family get-togethers, even in the sitting room, interrupting himself; often brusquely in the midst of a conversation.

Although Poincaré had a general grasp of all branches of mathematics, he was apparently ignorant of much of the literature. One consequence of this was that each new subject he heard about drove his interests in a new direction. Thus, when he heard about the work of Riemann and Weierstrass on Abelian functions, he threw himself into the theory of Abelian functions. Finding himself under a constant influx of ideas from the most diverse fields of mathematics, Poincaré did not have time to be rigorous, it was said; he was often satisfied when his superb intuition gave him the confidence that the proof of such and such a theorem could be carried through to complete logical rigor and then left the completion of the proof to others. Sometimes this led him astray, as on the following occasion.

King Oscar II of Sweden and Norway was interested in mathematics. To mark his sixtieth birthday a special prize competition was announced for an important discovery in certain areas of mathematical analysis. It was hoped that Poincaré would compete. His entry, submitted anonymously in accordance with the rules but easily identifiable by its style and command of the subject, was awarded the prize, and preparations were made for its publication. However, Poincaré discovered serious mistakes in his memoir at a late stage, which had to be corrected at his expense. This cost Poincaré more than the value of the prize he received, quite apart from the resulting embarrassment. Eventually the corrected memoir was to form the basis for his masterpiece, the *Methodes nouvelles de la mécanique céleste*.

Although Poincaré was said to be incapable of understanding the simplest principle of administration, he was active on various scientific committees. When secretary of the Société Mathématique de France, he proposed Georg Cantor for membership. While Poincaré was very enthusiastic about Cantor's early work, he objected strongly to his later axiomatic set theory. As David Hilbert said later, "regrettably Poincaré, who in his generation was the richest in ideas and the most fertile, had a decided prejudice against Cantor's theory that kept him from forming a just opinion of Cantor's magnificent conceptions" (Dauben 1979). Poincaré maintained good relations with the German mathematicians and visited Göttingen toward the end of his life to give some lectures, one of which was on the foundations of mathematics. Although he intended to be polite (he could be devastatingly impolite if he tried), Poincaré fulminated in this lecture against Cantor's theories. One member of the audience (Ernst Zermelo) was so offended he threatened to shoot Poincaré.

The self-reporting of experiences, particularly long after the fact, may not always be reliable. Fortunately, in the case of Poincaré we have some independent testimony, both in the Poincaré archives and in the remarkable account of his working methods written by the psychologist Edouard Toulouse, who was particularly interested to compare Poincaré's methods with those of the writer Emile Zola. Toulouse was director of the laboratory of experimental psychology at the École des Hautes Études in Paris, where he and his associates conducted a "medical-psychological" enquiry into "intellectual superiority." He found Poincaré, whom he examined in 1897, a fascinating subject.

Toulouse started by recording Poincaré's physical statistics at the age of 53. He was 1.65 m in height and 70 kg in weight. His hair was a chestnut color; he had a large nose and a blond moustache. Another observer, a few years later, described him rather unkindly as a dwarfish man with a slightly hunched back, a rough short beard, and very sad eyes. His daily routine was to rise at 7 am, breakfast at 8, lunch at noon, dine at 7, and go to bed at 10. Often troubled by indigestion, he did not sleep well.

In spite of his poor eyesight, Poincaré's principal diversion was reading. A book once read—at incredible speed—became imprinted on his mind so that he could always state the page and line where a particular item occurred. He retained this photographic memory all his life. His temporal memory—the ability to recall with precision a sequence of events long past—was also unusually strong. He saw the letters of the alphabet in color.

Toulouse knew that the thought of most mathematicians is visual, and was surprised to find that Poincaré was an auditory thinker, who apparently neglected visual images altogether. For example, when shown a table of letters or numbers, Poincaré repeated the table to himself, rather than trying to recall its visual image. Generally he found memorization difficult and was poor at rote learning, instead seeking patterns. Some people have no problem memorizing large amounts of material; others have trouble where there are no clear patterns. Given a scientific problem, many scientists attempt to recall previous work on it, while others work from first principles, like Poincaré.

Poincaré depended on unconscious thought for many of his key discoveries. Toulouse concluded that Poincaré's problem-solving methods were intuitive, rapid, and spontaneous. He used no notes to write his papers. He would have no definite overall plan or goal in mind, nor even any idea of whether the problem at hand was solvable. Getting started was never difficult, and after that he seemed led by his work without the impression of any willful effort. He often wrote down a formula mechanically in order to awaken some association of ideas. If progress became difficult, he did not persist but turned to something else. He proceeded by sudden bursts, first stopping and then starting again after his unconscious had done the work. At that stage it was difficult to distract him; he often found it hard to cease work. For this reason he never continued working on anything important in the evening so as not to disturb his sleep.

Toulouse believed that Poincaré knew just how to activate areas of long-term memory, and that this method of work accounted for his extraordinary creativity. In his dream-like approach to research, he opened up the boundaries of thought beyond the limits imposed by conscious deliberations. To make geometry, Toulouse wrote, or to make any science, something other than pure logic is necessary. To describe this something, we have no word other than "intuition." Poincaré severely criticized mathematical philosophers for neglecting the psychological origin of mathematical concepts. Other mathematicians have recorded the role of the unconscious in their creative work, but Poincaré's complete confidence in the power of his unconscious mind seems unparalleled.

Apart from diphtheria in early childhood and an attack of rheumatoid arthritis at the age of 32, Poincaré enjoyed good health until 1908, when a prostate enlargement began to give cause for concern. In 1908 at the fourth International Congress of Mathematicians in Rome, those present were greatly worried when he needed to return to Paris for medical treatment before he was able to deliver his scheduled address on the future of mathematics. In 1911 he took the unusual step of sending an unfinished paper to be published, afraid that he might not live to complete it. He died on 17 July 1912, at the age of 58, due to a postoperative embolism. The mathematical world was deeply shocked at the untimely death of the greatest mathematician of his time, and one of the greatest of all time.

Poincaré had many of the characteristic features of high-functioning autism, a disorder similar to Asperger syndrome (see Fitzgerald [2004] for a discussion of the differences). His fellow mathematician Paul Appell recalled that at the lycée he appeared "absorbed in his inner thoughts . . . when he spoke his eyes were filled with an expression of kindness, at the same time malicious and profound. I was struck by the way he talked in brief, jerky sentences, interspersed with long silences" (James 2002). Others noted, "When speaking to him one has the feeling that he had not followed or understood what was said, yet he answers or thinks about it," and, "He is neither sociable nor ready for confidences. He speaks correctly, but with some shyness." It was also said that "in practical life he is disciplined but absent-minded, for example he frequently forgot his meals and almost never could remember whether or not he had breakfasted" (James 2002).

Raymond Poincaré, who became prime minister of France and then president of the French Republic during the First World War, also had many of the characteristics of high-functioning autism. Raymond and Henri were first cousins, their fathers being brothers, which points to the disorder coming from that side of the family. On the other hand, Henri traced his mathematical ability to his maternal grandmother, who found mental arithmetic very easy.

David Hilbert

David Hilbert was born on 23 January 1862 in Wehlau, near Königsberg. On his father's side his more immediate forebears were professional men who practiced in the capital and coronation city of the Prussian kings. His father

Otto was a county judge; his mother Maria Theresa (née Erdtmann) was interested in mathematics, and Hilbert is said to have inherited his gifts from her side of the family. The future mathematician was their only son, but he had a younger sister who died in childbirth at the age of 28. In the family David was regarded as "a bit off his head"; he described himself as "dull and silly." His father was described as being rather narrow in his views, with strict ideas about proper behavior, and so set in his ways that he took exactly the same walk every day.

When Otto Hilbert was promoted to a senior judgeship in Königsberg, the family moved into the city, where his son was enrolled at the Royal Friedrichskolleg at the age of 8. The curriculum at this traditional grammar school emphasized the classics; no science was taught, apart from a little mathematics. It was hardly the most suitable Königsberg school for someone like David Hilbert, and those who would later become close friends of his attended the more progressive Wilhelms-Gymnasium. He only began to display his true abilities when he completed his school career with a year at the Gymnasium followed by three years studying for a doctorate at the University of Königsberg, apart from two semesters at Heidelberg.

Hilbert passed the doctoral examination at the end of 1884 and embarked upon the customary tour of other centers. He went first to visit Felix Klein in Leipzig, thus beginning a long-lasting collaboration, and then went on to Paris, where he called on Poincaré, among others. On the way back to Königsberg he stopped in Berlin and met Leopold Kronecker, who had just succeeded Ernst Kummer at the university. By 1886 Hilbert was a *privatdozent* at Königsberg; he retained this position until in 1892 he was appointed associate professor. In the same year he married Käthe Jerosch, the independent-minded daughter of a local merchant. She described her husband as "a bit of an arrested juvenile." His wife was there, he believed, simply to look after him and provide the best possible home environment for creativity; without her, it was said, he could not have lived the life he did. Yet "she was a full human being in her own right, strong and clear, and always stood on the same footing with her husband, kindly and forthright, always original" (Reid 1970). She tolerated his tendency to flirt with other women. Their only child Franz, born in 1893, became mentally ill.

Although Hilbert was deeply attached to his hometown, he had ambitions to move to a place where there were better mathematics students than at Königsberg. The Hilberts used to read the newspaper at breakfast each morning just for news about the health of professors of mathematics. In 1895 the opportunity he had been waiting for came up. Kronecker died, freeing a chair in Berlin and leading to a cascade of professorial appointments at German universities. At the end of this process Klein had secured Hilbert for the Georgia Augusta, the university with which he was to be closely identified and where he remained for the rest of his professional career. At the same time, Hilbert succeeded Klein as principal editor of the *Mathematische Annalen*. According to Richard Courant (1981), "He took this very seriously and the violence with which he rejected papers was completely without sympathy." Hilbert's directness was "something to be afraid of."

The impression Hilbert made when he arrived in Göttingen is described by Courant:

> In the Göttingen society, if you read old chronicles, a Göttingen professor was a demi-god and very rank conscious—the professor, and particularly the wife of the professor. Hilbert came to Göttingen and it was very, very upsetting. Some of the older professors' wives met and said "Have you heard about this new mathematician who has come? He is upsetting the whole situation here. I learned that the other night he was seen in some restaurant, playing billiards in the backroom with privatdozents." It was considered completely impossible for a full professor to lower himself to be personally friendly with younger people. But Hilbert broke this tradition completely, and this was an enormous step towards creating scientific life; young students came to his house and had tea or dinner with him. Frau Hilbert gave big, lavish dinner-parties for assistants, students, etc. Hilbert went with his students and also everybody else who wanted to come, for hour-long hikes in the woods during which mathematics, politics and economics were discussed.

Hilbert's mathematical interests were remarkably broad. As Courant puts it: "The most impressive thing was the great variety, the wide spectrum, of his interests. Enormously important to Hilbert throughout his life was the variety in all aspects of mathematics." He would make a decisive impact on some branch of mathematics and then move on to something quite different. Courant describes Hilbert's method of dealing with a problem as follows:

> He was a most concrete, intuitive mathematician who invented, and very consciously used, a principle: namely, if you want to solve a problem first strip the

problem of everything that is not essential. Simplify it, specialize it as much as you can without sacrificing its core. Thus it becomes simple, as simple as it can be made, without losing any of its punch, and then you solve it. The generalization is a triviality which you do not need to pay too much attention to. This principle of Hilbert's proved extremely useful for him and also for others who learned it from him; unfortunately it has been forgotten.

In 1902 Hermann Minkowski, one of his mentors at Königsberg, came to join Hilbert at the Georgia Augusta. Although the two mathematicians were very different in personality and background, they stimulated each other intellectually, and Minkowski provided the companionship Hilbert needed and could not find in Göttingen. "The conversation of the two friends was an intellectual fireworks display," said one of their companions, "full of wit and humour and still also of deep seriousness" (James 2002). For Minkowski, as for Hilbert, number theory was the most wonderful creation of the human mind and spirit, equally a science and the greatest of arts. However, they could not resist trying to place the new developments in fundamental physics on more satisfactory mathematical foundations.

In the summer of 1908, Hilbert became very nervous and depressed. "Almost every great scientist I have known," said Courant, "has been subject to such deep depressions. There are periods in the life of a productive person when he appears to himself and perhaps actually is losing his powers." However, after several months of rest in a sanatorium, Hilbert was back to work. The next year he was deeply shocked by the sudden death of Minkowski. A few years later there was a further blow when he learnt that his mentally ill son Franz was likely never to recover: from now on, he is reported to have said, I must consider myself as not having a son. As an adult Franz was often a patient in a psychiatric hospital. He often expressed himself inappropriately, and he wanted to save his parents from the evil spirits he believed were after them. When he returned home from time to time, the peace of the Hilbert household was disrupted. It is said that he bore an extraordinarily close resemblance to his father in appearance. There is a parallel with Einstein here; he too rejected a mentally ill son whose resemblance to his father was described as quite disconcerting.

Hilbert's wide range of interests had its drawbacks. According to Hermann Weyl, who attended his undergraduate lectures around 1910, Hilbert tended to move from one theory to another and from one discipline to the next without providing motivations, without explanations of historical background, without giving explicit references to his sources, and without stopping to work out any particular ideas or prove any assertion in detail, claiming all the while to possess a unified view of such matters. Hilbert usually avoided reading the literature, preferring to get his information from lectures and conversation. Although Weyl was deeply impressed by Hilbert, he commented that the understanding Hilbert imparted to students did not run very deep. This may be contrasted with Courant's (1981) description of Hilbert's lecturing style:

> His lectures were not perfect in a formal way, and it happened quite often that he had not prepared quite enough, so that at the end of the hour he would run out of material and had to improvise, which made him stumble and fumble. His friends and students made fun of him and gave him all kinds of ironical gifts for his birthday, to help him to stretch the content. He also made mistakes and got stuck in proofs, and so you had the chance to observe him struggling with sometimes very simple questions of mathematics, and finding his way out. This was more inspiring than a wonderfully perfect performance lecturing.

When the First World War broke out, the Georgia Augusta lost the majority of its students and some of its faculty members as well. Hilbert was absorbed in the foundations of physics, trying to place Einstein's revolutionary work on a more elegant mathematical foundation. Next he turned to the foundations of mathematics, where he led the opposition to Brouwer's theory of intuitionism and was dismayed when Weyl became a supporter of Brouwer. Hilbert's goal was to devise an axiomatic system that would place the foundations on a logical basis. Another of his early mentors, Adolf Hurwitz, died just after the war ended, and in 1925 he lost his eminent colleague Felix Klein.

In spite of the glow that surrounded Göttingen in the 1920s, there was a certain amount of hostility toward the faculty both from within Germany and from further afield. Partly this was jealousy, but there was also a tendency at the Georgia Augusta not to worry too much about what was happening elsewhere, and as a result a certain carelessness about attributions. For instance, Hilbert's celebrated *Zahlbericht* monograph on algebraic number theory made extensive use of an earlier survey by the Oxford mathematician Henry Smith. As Courant (1981) explains:

[Hilbert] was completely open, open to criticism and open to different points of view and every student. Everybody who had contact with him felt that although he was such a mental giant and such a really great force in science, one could talk to him on an equal footing—if one had something to talk about. Hilbert was not a scholar in the sense that he knew everything that happened in the world. He did not read every paper nor have a little catalogue in which he could find out everything that existed. On the contrary, it was one of his strengths, but also one of his shortcomings, that he listened very carefully and caught inspiration, but then frequently forgot whence the inspiration came.

Ever since the depressive episode of 1908, Hilbert's health had declined, but it was not until 1925 that his illness was diagnosed as pernicious anemia, a condition later found to be caused by a deficiency of vitamin B_{12}. The symptoms are both physical and psychological. Not long before, the mathemetician Sophus Lie had died from the condition, but Hilbert was able to benefit from a new treatment developed in the United States, although he never fully recovered. He retired in 1929 and ceased research, although he continued to lecture occasionally. A street in Göttingen was named after him, while Königsberg, his birthplace, made him an honorary citizen. In 1932, Göttingen celebrated his seventieth birthday with the traditional torchlight procession. Soon Hilbert began to experience memory loss and appeared to believe he was still living in Königsberg. After a period of senility, he died in Göttingen on 13 February 1943, ten years after the Nazis came to power. Under wartime conditions, barely a dozen people attended his funeral.

To those who knew him, Hilbert's presence was perhaps as influential as his work. Of his physical appearance in his prime, someone who had known him well wrote: "If I were a painter, I could draw Hilbert's portrait, so strongly have his features engraved themselves into my mind, forty years ago when he stood on the summit of his life. I still see the high forehead, the shining eyes looking firmly through the spectacles, the strong chin accentuated by the short beard, even the bold panama hat, and his sharp East Prussian voice still sounds in my ears" (Reid 1970).

In the words of Hilbert's disciple Hermann Weyl:

The impact of a scientist on his epoch is not directly proportional to the scientific weight of his research. To be sure, Hilbert's mathematical work is of great depth and universality, and yet his tremendous influence is not accounted for by that alone. Gauss and Riemann are certainly of no lesser stature than Hilbert but they made little stir among their contemporaries and no school of devoted followers formed around them. But Hilbert's nature was filled with a zest for living, seeking intercourse with other people, above all with younger scientists, and delighting in the exchange of ideas. His optimism, his spiritual passion, his unshakeable faith in the supreme value of science, and his firm confidence in the power of reason to find simple and clear answers to simple and clear questions were irresistibly contagious. (Reid 1970)

In relation to mathematics, Hilbert showed enormous self-discipline and self-control. He was equally controlling of others: a colleague told him, "You have made us all think only that which you would have us think" (Reid 1970). Then there was the peculiar stiff gaze, the shining eyes, the unconventional dress, and much else. While these are suggestive of Asperger syndrome, what Weyl says of his nature makes this seem doubtful.

An inspiring teacher as well as a great researcher, Hilbert was arguably the leading figure in the mathematical world in the years following the death of Poincaré. The Nobel laureate Max von Laue described him as the greatest genius he had ever laid eyes on. Especially in the first quarter of the twentieth century, mathematicians flocked to Göttingen to hear him and fell under his spell. Through their reminiscences we can obtain some idea of Hilbert's extraordinary charisma. As his biographer Constance Reid (1970) observed, like all great original minds in mathematics, he had the naivety and the freedom from bias and tradition that are characteristic of a truly great investigator. Reid's excellent biography is the main source of material for this profile, but the quotations are mainly taken from the reminiscences collected by Richard Courant (1981).

Chapter 7 } Hadamard, Hardy, Noether, and Ramanujan

Jacques Hadamard

Jacques Hadamard was born in Versailles, southwest of Paris, on 8 December 1865. Most of his forebears on both sides were intellectuals of Jewish extraction, and had been based in Paris since 1808. Hadamard's father taught Latin at the Lycée Charlemagne, and his mother was a noted music teacher who taught her son to play the violin at an early age. In 1871, returning from exile at the end of the Paris Commune, they found that their house had been burned down and they had lost all their possessions.

After recovering from this setback, they sent their son to the lycée where his father taught; he shone in every subject except mathematics, which he did not care for. Although he won prizes in other subjects in the national Concours Généraux, at first he showed no mathematical ability whatsoever. In 1875, when his father was transferred to the more prestigious Lycée Louis-le-Grand, the young Jacques followed him there and experienced mathematics teaching of high quality. When he took the Concours Général in mathematics again, he placed second in the whole of France: throughout his life he felt ashamed of his failure to take first place. He then sat the entrance examinations for the École Polytechnique and the École Normale Supérieure. In 1884, at the age of 18, Hadamard came first, in the whole of France, in both examinations. He chose the Normale.

After graduating in 1888, Hadamard spent a further year studying in Paris, supported by a sinecure teaching post at Caen. He then taught for three years at the Lycée Buffon, where he was not a success. "M. Hadamard believes himself exempted from everything because of his remarkable mathematical abilities," reported the headmaster to the minister of education. After five years on the faculty of the University of Bordeaux, he returned to Paris as lecturer at the Sorbonne and deputy professor at the Collège de France, and from then on Hadamard's career was one of unbroken success. By 1912 he was also professor of analysis at the École Polytechnique. When the Academy announced a prize competition, it was often Hadamard that won it.

The death of Poincaré left a vacancy at the Paris Academy to which Hadamard, who had been candidate several times previously, was finally elected at the age of 47. He was one of Poincaré's greatest admirers; when Poincaré died Hadamard put everything else aside to write on the life and work of the friend and colleague he described as the supreme genius.

Hadamard married a childhood sweetheart with a similar background to his own. She was a vivacious young woman named Louise who shared his love of music. They had five children: Pierre in 1894, Etienne in 1897, Mathieu in 1899, Cecile in 1901, and Jacqueline in 1902. Jacques and Louise made sure all of the children learned to play at least one musical instrument.

Hadamard mainly worked at home. His daughter Jacqueline gave a picture of her father at work:

> He practically never wrote a word. He always told me that he thought without words, and that for him the greatest difficulty was to translate his thoughts into words. He only scribbled down equations, not at a table but at a high wooden plinth of the kind that were normally used at that time to put a bust on (at my grandmother's it was a bust of Beethoven, of course). In the hall he would write down his mathematical formulas while walking up and down and for many years I used to hear my mother taking down dictated sentences of the kind "we integrate-poum we see that the equation-poum-poum-poum" (the number of poums indicated the length of the space to be left for the formulas).

The Hadamards kept open house for guests. The years before the First World War were an exceptionally happy time for Jacques: new theorems, lectures, a loving wife and children, friends, and a comfortable house filled with the sound of music. At one stage the Hadamards, in collaboration with some other like-minded academics, educated their children at home. Later the sons attended their father's old school, the Lycée Louis-le-Grand. Pierre, the eldest, had just been accepted for the École Polytechnique when the war came, while Etienne, the second eldest, was accepted for the École Normale two years later. Both were killed in action on the western front. Toward the end of the war, the Hadamards' youngest son, Mathieu, was called up but posted to the less dangerous Italian front and survived. Jacques Hadamard was involved in military research, while Louise served as a nurse.

When Norbert Wiener first went to Paris as a postdoctoral student, he found that usually after a professor had given a lecture he retired to his little office to sign the daybook that gave a record of his lecture and then vanished from the lives of his students and younger colleagues. Hadamard was an exception; he was genuinely interested in his students and always accessible to them:

> He has considered it an important part of his duty to promote their careers. Under his personal influence the present generation of French mathematicians, for all the tradition of a barrier between the younger and the older men, has gone far to break down this barrier. I myself benefited from Hadamard's largeness of outlook. There was no reason why Hadamard should have paid any particular attention to a barbarian from across the Atlantic at the beginning of his career. That is, there was no reason except Hadamard's good nature and his desire to uncover mathematical ability wherever he could find a hint of it.
> (Wiener 1953)

In 1921 Hadamard inaugurated the twice-weekly seminar that rapidly became famous and ran for over twenty years. Initially it was the only serious window on international mathematics available in Paris. It was considered a great compliment to be invited to lecture in it. Its success was due not only to Hadamard's exceptional mathematical insight and wide interests but also to his cordiality, humanity, and sense of humor.

Hadamard enjoyed mountaineering and was proud of having climbed Mont Blanc when over 60 years of age. He had an untiring instinct for justice. The Dreyfus affair, in which a Jewish army officer was unjustly convicted of espionage, was an obsession with him, not simply because the victim was a distant relative. It led him to become a communist sympathizer and an avowed pacifist. As well as his fascination with the psychology of mathematical invention (following in the footsteps of Poincaré), Hadamard had a lifelong interest in botany, especially fungi and ferns. He loved traveling, and hardly a year went by without a visit to another country for one reason or another. When he crossed the Soviet Union on the trans-Siberian railway, he left the train at each stop in search of botanical specimens to add to his collection, to the alarm of Louise, who feared he would be left behind when the train started again.

After the fall of France in 1940, the Hadamards escaped to the United States. Two years later they moved to England, where they stayed until the war was over. When they were able to return home to France to live out the rest of their lives quietly, they found that the Germans had ransacked their apartment. The family papers were never recovered. Louise died in 1960, after sixty-eight years of marriage. Hadamard's own health declined; he suffered hearing loss and experienced difficulty in walking. He died on 17 October 1963, at the age of 97. By then, he had received almost every academic honor France had to offer and many foreign honors as well.

Hadamard was an outstanding all-rounder, just as strong in teaching as in research, and extraordinarily versatile. His scientific curiosity was insatiable. During the course of his long life mathematics changed enormously, but he kept abreast of it. As a schoolboy, André Weil met Hadamard, of whom he said later, "The warmth with which he received me eliminated all distance between us. He seemed to me like a peer, infinitely more knowledgeable, but hardly any older. All those who were acquainted with him know that until the end of his very long life, he retained an extraordinary freshness of mind and character: in many respects his reactions were those of a fourteen-year-old boy. His kindness knew no bounds" (Weil 1992).

G. H. Hardy

Perhaps surprisingly, there is no full-scale biography of Hardy, although we have his famous memoir (1940/1967), which is partly autobiographical. This is the main source for the following profile, but the relevant chapter of Kanigel (1991) has also been used.

Godfrey Harold Hardy was born on 7 February 1877 in the village of Cranleigh, near Guildford, in the county of Surrey. He and his younger sister Gertrude were the children of schoolteachers. Their father taught geography and drawing at Cranleigh School, where he also gave singing lessons, edited the school magazine, and coached football. He is described as gentle, indulgent, and somewhat ineffectual. Their mother, who gave piano lessons and helped run a boarding house for the younger pupils of the school, was rather the opposite. She is described as obsessively involved with her children's welfare and education.

Hardy displayed a special talent and interest in mathematics from an early age. He passed the time at church by factorizing the numbers of the hymns. Rather than attend regular classes in mathematics, he was coached privately and reached the top form at Cranleigh when he was only 13, ranking second in the class. He then won a scholarship to Winchester, one of the leading boarding schools, where he spent six unhappy years before leaving in 1896 to go to Trinity College, Cambridge, as a scholar. He never forgave Winchester for not giving him the opportunity to play cricket for the school.

At Cambridge, the examination system was desperately in need of reform. Hardy considered the mathematical Tripos examination an utter waste of time and tried to change his course from mathematics to history. However, after reading Camille Jordan's *Cours d'analyse* he learned for the first time what mathematics really meant. "From that time onwards," he wrote, "I was in my way a real mathematician with sound mathematical ambitions and a genuine passion for mathematics."

After graduating in 1901, he was elected to a prize fellowship at Trinity; the following year he won the coveted Smith's Prize. As a fellow, Hardy was finally free to devote his time to pure mathematics, and he did so with great enthusiasm and fervor. With others of a similar persuasion, he set about trying to reform the mathematical Tripos and managed to have some of its worst features eliminated. After his fellowship came to an end, he was appointed to a college lectureship, with an obligation to give six lectures a week.

In 1910 Hardy was elected to the Royal Society of London. Despite what many saw as a highly productive period between the years 1900 and 1910, he himself felt that he did not publish much of real value in that decade: "I wrote a great deal in the next ten years but very little of any importance; there are not more than four or five papers I can remember with some satisfaction" (Hardy 1940/1967). He believed his best work came afterwards, out of his associations with J. E. Littlewood and S. Ramanujan.

Hardy's collaboration with Littlewood, who was eight years his junior, began in 1911. The partnership, which lasted over thirty-five years and resulted in the publication of over a hundred papers, was described by C. P. Snow as "the most famous collaboration in the history of mathematics. There has never been anything like it in any science or in any other field of creative activity" (Hardy 1940/1967). The two men worked together mainly by correspondence, even when they were living in the same college. At meetings and conferences it was always Hardy who presented their joint work. Indeed, Littlewood so seldom attended on these occasions that mathematicians used to amuse themselves by speculating as to whether Littlewood really existed or was simply a figment of Hardy's imagination.

During Hardy's early years at Cambridge, he was part of several social groups, including the select essay club known as the Apostles. This met weekly to discuss and debate philosophical issues, and over the years boasted some brilliant minds among its membership. Some of Hardy's closest friendships were formed in it. During the First World War, when Littlewood was away serving in the army, Hardy found his duties becoming increasingly administrative, interfering with his research. When the Savilian Professorship of Geometry at Oxford became vacant in 1919, Hardy took the opportunity to leave Cambridge. By virtue of his chair he became a professorial fellow of New College, which has historic links with Winchester, his old school. At Oxford he reached the peak of his career, setting up a flourishing research school and enjoying the best years of his collaboration with Littlewood, who remained in Cambridge. At Oxford, "mathematics is one of the traditional subjects of study, but it probably attracts less notice than in any other considerable university," he wrote. "Mathematics and physics are overshadowed not merely by the literary schools but even by the other sciences" (Hardy 1940/1967).

On social occasions Hardy was an entertaining talker on a wide variety of topics. Conversation was a game he liked to play, and it was not always easy to make out what his real opinions were. Unusually sensitive and self-conscious, like his sister, he concealed an underlying seriousness by posing as a mildly eccentric don. He would never carry a watch or use the telephone, corresponding mainly by telegrams and postcards. Some of Hardy's eccentricities send a message to the psychologist. For example, he could not abide mirrors—he would not have them in his own rooms, and if he stayed anywhere else he would immediately cover up any mirrors. The reason for this strange but not uncommon behavior is thought to be a fear of losing control—an anxiety that the person seen in the looking glass is going to capture the identity of the person looking into it.

"I suspect," said someone who knew him later in life, "that Hardy found many forms of contact with life very painful and that, from a very early stage, he had taken extensive measures to guard himself against them. His formidable wall of charm and wit shielded an immensely fragile ego. Certainly in his friendlier moments—and he could be very friendly indeed—one was conscious of immense reserves" (Kanigel 1991). Always he kept the world at bay. The obsession with cricket, the bright conversation, the studied eccentricity, the fierce devotion to mathematics—all of these made for a beguiling public persona, but none encouraged real closeness. Hardy did not normally form close, demonstrative bonds, even with those he called his friends. He kept the various parts of his life scrupulously walled off from each other.

In 1931 Hardy returned to Cambridge as Sadlerian Professor of Pure Mathematics and was elected a fellow of Trinity again. One of the reasons he gave for the move was that eventually, when he reached the mandatory retirement age, he would be able to continue to live in college, something not normally permissible at Oxford. However, as an Honorary Fellow of New College, he was able to retain his connection with Oxford and frequently spent his weekends and vacations there. Hardy had an exceptional gift for working well with other people, and besides Littlewood and Ramanujan, he collaborated with many other leading mathematicians of the day. Another side to his nature showed in his efforts to promote reconciliation with the German mathematicians after the First World War and, later, to help resettle the academic refugees from Nazi-occupied Europe.

The London Mathematical Society occupied an important place in Hardy's life. He served on the council from 1905 to 1908 and from 1914 almost continuously until his final retirement in 1945. He was one of the secretaries from 1917 to 1945 and twice president, never missing a meeting. Although no great traveler, Hardy made several visits to the United States and had many friends there. He spent 1928–1929 as exchange professor at Princeton and the California Institute of Technology. Although Hardy was highly regarded in his home country, and the recipient of many honors, he was perhaps more appreciated overseas, particularly in Germany. Hilbert thought him the best mathematician in England, particularly when he was collaborating with Littlewood.

George Pólya, who knew him well, said "Hardy had quite a special personal charm. I cannot describe it, what the charm was, but everyone was attracted to him, men and women, mathematicians and non-mathematicians, intellectuals and very simple people." In fact there was something rather boyish about him; that may have been part of the attraction, but some people felt he carried it too far. For example, he pretended to think that God was a personal enemy. When going on a sea voyage, for example, he would leave behind a postcard addressed to one of his friends that read, "Have just proved the Riemann hypothesis." The idea was that if he did not return, Hardy would go down in history as the man who proved the hypothesis (method unknown). Since God would not want him to obtain this posthumous fame, he would ensure that Hardy returned safely; the postcard would then be destroyed.

Hardy was a consummate lecturer: his enthusiasm and delight in the subject fairly spilled over. "In all my years of listening to lectures on mathematical analysis," said Norbert Wiener, "I have never heard the equal of Hardy for clarity, for interest or for intellectual power" (Wiener 1953). But lucid as Hardy's lectures were, his writing probably had greater influence. He wrote, in his own clear and unadorned fashion, some of the most perfect English of his time. As well as a large number of mathematical papers, some on his own, others in collaboration, he wrote several mathematical books that rank as classics, and he also wrote several nontechnical books. His obituaries of famous mathematicians were rounded, gracious, and wise. Journalism was the only profession outside academic life, he said, in which he would have felt really confident of his chances.

Outside mathematics, Hardy's main passion was the game of cricket, which he loved to watch, talk about, and occasionally play. He was inclined to form intense romantic attachments to young men. These were intense affections, absorbing, nonphysical, but exalted. "One I knew about," wrote Snow in his foreword to the 1967 edition of Hardy's memoir, "was for a young man whose nature was as spiritually delicate as his own. I believe, although I only picked this up from chance remarks, that the same was true of the others." Hardy owed much to his devoted sister, who was also gifted mathematically and of a similar personality in many ways. In 1939 he suffered a coronary thrombosis, which put an end to the sporting activities, such as cricket and real (or royal) tennis, that he enjoyed so much. In his last decade he made an attempt to take his own life. Feeling his mathematical powers waning, he began in 1940 to write his testament, *A Mathematician's Apology* (1940), from which we have been quoting. This jewel of a book is still in print and has been translated into many languages. When Hardy died, on 1 December 1947, in his seventy-first year, he was listening to his sister read to him from a book about the history of cricket at Cambridge.

Hardy certainly had Asperger traits. Painfully shy and self-conscious, he was born with three skins too few. He disliked small talk, never made eye contact with other people; always there was a haunted look in his eyes. He abhorred formal introductions, would not shake hands, would walk along the street with his face turned down, ignoring those who might expect him to exchange greetings. "My devotion to mathematics is indeed of the most extravagant and fanatical kind," he wrote. "I believe in it, I live it, and should be utterly miserable without it. My mathematical research," he would say, "was the one great permanent happiness of my life" (Hardy 1940/1967).

Emmy Noether

Amalie Emmy Noether was born on 23 March 1882 in Erlangen, Germany, where her father was professor of mathematics at the university. Her mother came from a wealthy Jewish family in Cologne. Noether grew up in a home regularly visited by members of a stimulating circle of scholars. It is said that her father strongly influenced the early thinking of his children. Otherwise she had a conventional upbringing, attending the municipal school for the education of daughters until she was 18, when she was certified as a teacher in French and English at "institutions for the education and instruction of females."

Emmy Noether was determined to continue her education at university level, and she was not easily deterred. After auditing courses in mathematics and other scientific disciplines at Erlangen and Göttingen, she matriculated at the former university in 1904, the first year it was possible for women to do so. Previously, the academic senate had passed a resolution declaring that the admission of women would overthrow all academic order. Her true abilities were quite slow to show themselves, but by 1908 she had completed her dissertation and was awarded a Ph.D. *summa cum laude.* For the next seven years she remained at Erlangen without a position, doing a little teaching but mainly engaged in research on algebra, where her work impressed David Hilbert. He tried to secure an academic position for her at the Georgia Augusta, but there was too much opposition. This provoked his famous outburst at a faculty meeting: "I do not see that the sex of the candidate is an argument against her admission as *privatdozent.* After all, we are a university not a bathing establishment" (Reid 1970). However, the opposition may have been as much because of her socialist political convictions as because of her sex. It was not until 1919 that she became even a *privatdozent,* at the age of 37. Three years later she received the honorary title of unofficial associate professor, without official responsibilities and without stipend, although she acted as thesis adviser for a number of Ph.D. students.

A keen mind and infectious enthusiasm for mathematical research made Emmy Noether an effective teacher. Her classroom technique, like her thinking, was strongly conceptual. Rather than simply lecturing, she conducted discussion sessions in which she would explore a topic with her students. On one famous occasion her slip came down when she was lecturing—she bent down, pulled it off, threw it in the corridor, and kept on lecturing. She kept a handkerchief tucked under her blouse: while lecturing she had a way of jerking it out and thrusting it back very energetically, which was very noticeable to her audience. Before she began keeping her hair cropped short, she wore it up and, little by little, in the excitement of lecturing, it would fall out of place. Every day she took her lunch at the same table in the same restaurant at the same time, eating the same plain meal.

Outstanding mathematicians often make their greatest contributions early in their careers. Emmy Noether was an exception: she began to produce her most powerful and creative work around the age of 40. At the Georgia Augusta she was at last appointed associate professor. She never attained the top rank of full professor, although she contributed so much to making Göttingen the premier mathematical center in Europe—many would say in the world. When the Nazis seized power in 1932, one of their first acts was to deprive non-Aryan government officials, including university teachers, of their positions, with certain exceptions. Most members of the mathematics faculty at the Georgia Augusta were Jewish; they were forbidden to teach at the university or even to enter the department. For a time Emmy Noether continued to meet informally with students and colleagues, inviting groups to her apartment while trying, with her despondent colleagues, to decide what to do. She considered moving to Moscow, where she had a following, but the authorities there were slow to act. In the meantime, efforts were being made on her behalf in the United States, and she secured a temporary position at Bryn Mawr College, the new college for women near Philadelphia.

Bryn Mawr offered fellowships for graduate study in mathematics. When

Emmy Noether arrived, there were five graduate students in the subject. She took four of them under her wing and taught them abstract algebra in a mixture of German and English (she did her best to speak English from the first). With characteristic curiosity and good nature she settled happily into her new home. She wanted to know how things were done in the United States, whether it was giving a tea party or taking a Ph.D., and she attacked each subject with the disarming candor and vigorous attention that won over everyone who knew her. As Hermann Weyl said, in his memorial address:

> Her work was as inevitable and natural as breathing, a background for living taken for granted, but that work was only the core of her relation with her students. She lived with them and for them in a perfectly unselfconscious way. She looked on the world with direct friendliness and unfeigned interest, and she wanted them to do the same, she loved to take walks, and many a Saturday, with five or six students, she tramped the roads around the college with a fine disregard for bad weather.

So what was she like personally? "Warm like a loaf of bread," said Hermann Weyl, adding that "there radiated from her a broad, comforting, vital warmth. She was strongly myopic and wore spectacles with thick lenses. She was fat, rough, and loud, but so kind that all who knew her loved her. She thought little about what she should wear, what she should eat, and so on; her intentions could hardly have been further removed from the effects that her appearance had, especially on those who did not know her."

At the end of that first year in the United States, she returned briefly to Germany and was appalled to find how much worse the situation had become. Toward the end of her second year at Bryn Mawr she went into hospital for surgery to remove a uterine tumor; although the operation was not without risk, it was carried out successfully. While in the hospital, she developed a high fever and died suddenly on 14 April 1935, at the age of 53. Shortly before she died, she remarked that the previous eighteen months had been the very happiest in her whole life, for she was appreciated in America as she never had been in her homeland.

Weyl concluded his memorial address with these words:

> It was only too easy for those who met her for the first time or had no feeling for her creative power to consider her queer or to make fun at her expense. She was heavy of build and loud of voice, and it was not easy to get the floor in competi-

tion with her. She preached mightily, and not as the scribes. She was a rough and simple soul, but her heart was in the right place. Her frankness was never offensive in the least degree. In everyday life she was most unassuming and utterly unselfish, she had a kind and friendly nature. Nevertheless she enjoyed the recognition paid to her, she could answer with a bashful smile like a young girl to whom one had whispered a compliment. No-one could contend that the Graces had stood by her cradle, but if we in Göttingen often chaffingly referred to her as *der Noether* with the masculine article, it was also done with a respectful recognition of her power as a creative thinker who seemed to have broken through the barrier of sex. She possessed a rare humour and a sense of sociability; tea in her apartment could be most enjoyable. She was a kind and courageous being, ready to help and capable of the deepest loyalty and affection.

Two traits determined her nature; first the native productive powers of her mathematical genius. She was not clay, pressed by the artistic hands of God into a harmonious form, but rather a chunk of human primary rock into which he had blown his creative breath of life. Secondly her heart knew no malice; she did not believe in evil—indeed it never entered her mind that it could play a role among men. This was never more forcefully apparent to me than in the last stormy summer, that of 1933, which we spent together at Göttingen. A time of struggle like that one . . . draws people together; thus I have a particularly vivid recollection of those months. Emmy Noether, her courage, her frankness, her unconcern about her own fate, her conciliatory spirit, were in the midst of all the hatred and meanness, despair and sorrow surrounding us, a moral solace. The memory of her work in science and of her personality among her fellows will not soon pass away. She was a great mathematician, the greatest, I firmly believe, her sex has ever produced, and a great woman. (Weyl 1968)

Srinivasa Ramanujan

The sad story of the life of the Indian mathematician Ramanujan has been told many times. Fitzgerald (2004) gives a detailed account and concludes that Ramanujan would meet the criteria for Asperger syndrome. The indications are that this probably came from the mother's side. She possessed mathematical ability herself and believed that her son's exceptional powers were supernatural. The dominant parent, she came from a line of famous Sanskrit scholars.

Srinivasa Iyengar Ramanujan was born on 22 December 1887 in his mother's hometown of Erode and raised in the city of Kumbakonam in the

Tanjore district of the Madras province of India. The family was orthodox Brahman but poor. His "very quiet" father worked as a bookkeeper in a cloth-merchant's establishment. His mother, a shrewd, cultured, and above all deeply religious woman, often sang devotional songs with a group at a local temple to supplement the family income. Her son bore a strong physical resemblance to her and remained close to her after he grew up.

As a child, Ramanujan would roll on the ground in frustration if he did not get what he wanted. Sometimes he would take all the domestic utensils in the house and "line them up from one wall to the other" (Kanigel 1991). His parents had other children who died while he was still quite young. Later two brothers were born who lived to manhood, but they were so much younger than him that Ramanujan was effectively an only child. He suffered from smallpox when he was 2. At the age of 3 he scarcely spoke at all; his mother was very worried about this.

Ramanujan received all his early education in Kumbakonam, where he studied English at primary school so that he could attend the town's secondary school, where the teaching was in the English language. He preferred his own company and had no interest in team sports; he walked with his body pitched forward onto his toes. His mathematical talents became evident early; at 11 he was already challenging his mathematics teachers with questions they could not always answer. Seeing his interest in the subject, some college students lent him textbooks. By the time he was 13, he had mastered a popular textbook on trigonometry used by students much older than himself. His classmates described him as "someone off in the clouds with whom they could scarcely hope to communicate" (Kanigel 1991). In 1904 he graduated from school, winning a special prize in mathematics and a scholarship to attend college.

Shortly before this, Ramanujan came across a book called *A Synopsis of Elementary Results in Pure and Applied Mathematics,* written by a British mathematics coach named Carr in the 1880s. Without much in the way of explanation, it listed thousands of results, formulae, and equations. In Ramanujan, this unleashed a passion for mathematics so overwhelming that he studied it to the exclusion of all other subjects. As a result he began failing examinations, and his scholarship was revoked. When he started proving results that were not in the *Synopsis,* some of these were completely original. He recorded them in a notebook that he showed to people he thought might be interested. A classmate described how Ramanujan "would open his notebooks and explain to me intricate theorems and formulae without the least suspecting that they were beyond my understanding or knowledge." It was clear that "once Ramanujan was lost in mathematics, the other person was as good as gone" (Kanigel 1991).

Without a university degree, it was difficult for Ramanujan to find a suitable job, and for some years he was desperately poor, often relying on friends and family for support. Occasionally he would tutor students in mathematics but without great success, because he did not keep to the syllabus and the standard methods. In 1909 his mother found a bride for him named Janaki, some ten years younger than he was. When she was old enough to leave her parents and live with him, his mother forbade them to share a bed. Soon after the marriage he left home and traveled to Madras, the principal city of south India, in search of a livelihood; eventually a professor of mathematics named Rao at the prestigious Presidency College, impressed by the work Ramanujan showed him, provided financial support for a while. Later Rao described how Ramanujan appeared to him at the time:

> a short, uncouth figure, stout, unshaved, not overclean, with one conspicuous feature—shining eyes—walked in with a frayed notebook under his arm. He was miserably poor. He had run away from Kumbakonam to get leisure in Madras to pursue his studies. He never craved for any distinction. He wanted leisure; in other words, that simple food should be provided for him without exertion on his part and that he should be allowed to dream on. He opened his book and began to explain some of his discoveries. I saw quite at once that there was something out of the way; but my knowledge did not permit me to judge whether he was talking sense or nonsense. Suspending judgement I asked him to come over again and he did. And then he had gauged my ignorance and showed me some of his simpler results. These transcended existing books and I had no doubt he was a remarkable man. Then, step by step, he led me to elliptic integrals and hypergeometric series and at last his theory of divergent series not yet announced to the world converted me. I asked him what he wanted. He said he wanted a pittance to live on so that he could pursue his researches. (Kanigel 1991)

In 1912 Ramanujan finally obtained a poorly paid post as an accounts clerk at the Madras Port Trust. When not so occupied, he would squat on the front porch of his house, madly writing on a large slate across his lap, seemingly oblivious to the squeak of the hard slate pencil. For all the noisy activity in the street, he inhabited an island of serenity. His wife would later recall that before going to work in the morning and when he came home in the evening, he would work on mathematics. Sometimes he would stay up till 6 a.m., then sleep for two or three hours before going off to his office work. Those who knew him in those days described him as friendly and gregarious, always full of fun. Even his lack of social sensitivity conferred on him an innocence and sincerity, so that people could not help but like him.

In 1913 Ramanujan began writing to leading mathematicians in Cambridge about his discoveries. The first two he approached were unhelpful; the third was G. H. Hardy. Something about the letter, perhaps its very strangeness, intrigued Hardy, and he decided that it was worth a closer look. Later he would rank it as the most remarkable he had ever received. Ramanujan, after a single paragraph of introduction, plunged into formulae and theorems stated without proofs. Hardy's first reaction was to disregard it as the work of a crackpot, filled as it was with wild claims and bizarre theorems without any reasoning offered in support. He consulted his colleague J. E. Littlewood; after three hours of perusal the two men decided they were dealing with a genius, not a crackpot. Hardy then wrote back to Ramanujan asking for some proofs, but Ramanujan either could not or would not furnish them; instead he provided even more results without proofs.

With Littlewood's support, Hardy decided to try to lure Ramanujan to Cambridge. However, a very orthodox Brahman, Ramanujan knew he would lose caste by traveling overseas and so declined. Hardy did not give up. As an interim measure, a two-year fellowship was arranged for Ramanujan at the University of Madras. Eventually, with the help of Hardy's colleague E. H. Neville, who was visiting Madras in 1914, Ramanujan was persuaded to make the journey to Cambridge.

Once in Cambridge, Ramanujan, under the guidance of Littlewood and especially Hardy, developed rapidly. However, he found that living in college had its problems. "Till now I did not feel comfortable and I would often think why I had come here" (Kanigel 1991), he wrote soon after arrival. Ramanujan was very particular about food, for religious reasons, and insisted on preparing his own meals; he had left his wife behind in India, where her mother-in-law was ill-treating her. There were cultural differences to overcome: he complained of being cold in bed, until it was suggested that he try sleeping under the bedclothes rather than on top of them.

At Cambridge, Ramanujan would work for up to thirty hours at a time. Students taunted him because of his shyness; it seems that he was not particularly attuned to interpersonal nuance. Indian visitors made a point of calling on him. They were most impressed by his mysticism. One of them left this account of what Ramanujan was like:

> a fair complexion, a slightly pitted face, dreamy eyes with an absent-minded look about them, the average height and a body slightly inclined to be stout— such was Ramanujan, the great mathematician. His rooms in the outer court of Trinity College, Cambridge, were as unimposing as their occupant. Even during the short period I was privileged to stay with him, I had many an occasion to hear from his visitors who were evidently taken by surprise of the unimposing figure of the great genius and the scantily supplied shelves. "Are you the great mathematician?" Naturally the very prime object of the query used to shrink further into his chair on such occasions. But after the visitor left, Ramanujan used to ask me "Can you suggest the appropriate reply to the question and describe the dramatic pose to be assumed in giving that reply?"

Ramanujan had an enormous sensitivity to the slightest breath of public humiliation. Even before he left India he had a reputation for sudden "disappearances." On one occasion, in 1916, he was entertaining some Indian friends at his apartment in Cambridge. He was proud of his skill in cooking. His guests had several servings of the South Indian food he had prepared, and after they expressed their appreciation, he offered them yet more. When the ladies present politely declined, he felt so mortified that he got up without a word, called a taxi, and disappeared for a week to Oxford. Such disappearances, in the wake of an intolerable blow to their self-esteem, are not unusual in people with Asperger syndrome.

It was hardly surprising that a large portion of the results in Ramanujan's notebooks consisted of rediscoveries; he had never had any systematic training or access to a good library. To quote Hardy:

> What was to be done in the way of teaching him modern mathematics? The limitations of his knowledge were as startling as its profundity. He had only the vaguest idea of complex analysis. Most of the theorems in his notebooks were not proved but only made to seem plausible. His ideas of what constituted a mathematical proof were of the most shadowy description. Because he tended to collapse many steps in his mathematical work it was very difficult for other people to work out how he got his results. He had a very deep mathematical in-

sight and a knack for manipulating formulas, a delight in mathematical form for its own sake. (Hardy 1921)

According to Hardy, Ramanujan "combined a power of generalisation, a feeling for form, and a capacity for rapid modification of his hypotheses, that were often really startling, and made him, in his own peculiar field, without a rival in his day" (Hardy 1940/1967). During the five years Ramanujan was in England he had over twenty mathematical papers published, including several written in collaboration with Hardy.

In the spring of 1917, Ramanujan became seriously ill and was treated for several months for pulmonary tuberculosis; it has recently been suggested that what he was suffering from was probably hepatic amoebiasis (Young 1994). Unfortunately the remote location of the sanatorium, which meant that he was cut off from his friends, combined with the spartan regime and, above all, the lack of acceptable food, left Ramanujan deeply depressed. In 1918 he attempted suicide. Despite this, according to Hardy's colleague Neville, Ramanujan never doubted that he did well to come to England. Meanwhile Hardy, with Littlewood's support, did all he could to ensure that Ramanujan's genius and contributions to the theory of numbers were recognized.

In 1918 Ramanujan was elected to the Royal Society and to a Trinity fellowship. Before long, however, owing to concern over his state of health, arrangements were made for him to return to India. He briefly held a professorship at the University of Madras, although in poor health. His mother and his wife continued their endless quarrels. He resisted medical treatment and died in Chetput, near Madras, on 26 April 1920 at the age of 32. Until the end, he remained passionately devoted to mathematics. His mother never recovered from his death. His wife, with no children of her own, was swindled out of the money he left her and condemned to the wretched existence of so many widows in India until the celebration of the centennial of his birth drew attention to her plight. His mathematical reputation suffered something of an eclipse in the years immediately after his death, but more recently it has recovered, and in his homeland he is regarded a national hero.

Neville, who knew Ramanujan well, summed up his character in these words: "perfect in manners, simple in manner, resigned in trouble and unspoilt by renown, grateful to a fault and devoted beyond measure to his friends, Ramanujan was a lovable man as well as a great mathematician."

Hardy said, "I owe more to him than to anyone else in the world with one exception, and my association with him is the one romantic incident in my life" (Hardy 1940/1967). The source of Ramanujan's mathematical powers remains a mystery. It seems fairly clear that a phenomenal memory played some part in it, but precisely what books he had seen and what he had learned from them remains unknown: no one thought to ask him. Ramanujan himself attributed his exceptional powers to the family deity, the goddess Namagiri. A deeply religious man, he combined his passion with his faith, and once told a friend, "An equation has no meaning for me unless it expresses a thought of God."

Chapter 8 } Fisher, Wiener, Dirac, and Gödel

R. A. *Fisher*

Ronald Aylmer Fisher was born in the Finchley district of north London on 17 February 1890. He and his twin brother, who died in infancy, were the youngest of eight children. His father was a member of a well-known London firm of auctioneers. Although most of the members of his father's family were in business, one of his brothers placed high in the Cambridge mathematical Tripos and went into the clergy. Fisher's paternal grandfather was said to have been inclined towards a scientific career. One maternal uncle became a successful London solicitor noted for his social qualities, while another gave up excellent prospects at home to collect specimens of wild animals in Africa.

Fisher was a mathematical child prodigy. At school he had the disconcerting habit of producing the correct answer to a problem without showing how he arrived at it. In later years, others found his work difficult to follow and criticized him for inadequate proofs and the use of intuition. As with many mathematicians, Fisher's special ability showed at an early age, and love of the subject dominated his professional career. At Harrow School he was fortunate to have good teachers. He was physically tough—he was keen on running. He suffered from extreme myopia, common in people of genius, and was forbidden to work by artificial light. This led to an exceptional ability to solve mathematical problems in his head and also to a strong geometric sense.

Fisher went up to Cambridge in 1909 as a scholar of Caius College. After three years he graduated with a first-class degree and stayed on for post-graduate work in physics. At the same time, he developed a strong interest in biology. During the First World War, Fisher worked as a schoolteacher, replacing men called up for military service, but he was a poor teacher and a poor disciplinarian, lacking the resonance with students that a teacher needs to get a sympathetic hearing, and he failed to arouse curiosity in his subject. So he battered his head against a brick wall of boyish mischief and passive incomprehension. It was not until after the war that he found his true vocation at Rothamsted, the agricultural research station outside London, where he built up an international reputation in genetics and in 1930 published his seminal work, *The Genetical Theory of Natural Selection*.

In 1933 Fisher was appointed Galton Professor at University College, London, succeeding Carl Pearson, with whom he maintained a long feud. The Galton Laboratory offered facilities for the experimental breeding of animals that had not been available at Rothamsted. He developed a mouse colony and also bred snails, dogs, and even marsupials. He took over the editorship of the *Annals of Genetics* from Pearson; under his guidance it became a journal of importance in statistics. He also published his standard work, *The Design of Experiments*. In 1943 he returned to Cambridge as Arthur Balfour Professor of Genetics and built up a flourishing school of mathematical genetics. He was elected to a fellowship of Caius, his old college, where he became a legendary figure to generations of undergraduates.

Fisher was awarded many high academic and scientific honors and was knighted in 1952. After retirement from his Cambridge chair in 1957, he led a nomadic existence until he settled in Adelaide, Australia, where he found that the climate and intellectual atmosphere suited him. In his later years, he often affected Indian dress. He enjoyed excellent health throughout his life, until he developed bowel cancer and died in 1962 following surgery.

Most obituary notices tell us little about the personality of their subject. However, in some of the *Biographical Memoirs of the Royal Society of London*, usually published several years after the death of the subject, there is an attempt to break with this convention. The memoir of Fisher by Yates and Mather (1963) contains the following.

> Fisher had a likeable but difficult character. He had many friends and was
> charming and stimulating to work with, and excellent company. He liked good
> food and wine, which he found gave an agreeable background to intellectual
> discussion. In conversation he brought not only a vast store of knowledge but

also an independent mind of great vigour and penetration to bear on almost any subject. He constantly questioned conventional assumptions. He was an idealist committed to the establishment of truth and the advancement of mankind. To human affairs in general he had a benign and tolerant attitude, and did not presume to sit in judgement on the personal conduct of his fellow men. He was fond of children and enjoyed having them around even when he was working. In politics he was an outspoken and lasting opponent of Marxism. He attended church regularly, in the belief that this was a salutary and humbling activity, but he did not subscribe to religious dogma.

Fisher's eccentricities, though sometimes embarrassing, were for the most part a source of entertainment to his friends. His poor eyesight, though a constant source of frustration, never seriously hindered his work or his enjoyment of life. He liked the company of other scientists and was a familiar figure at scientific meetings. He was extremely generous with his ideas and always ready to discuss problems with others; he often came up with a solution with surprising speed.

These were among the positive and very great virtues of the man. On the negative side must be mentioned his notoriously contentious spirit, his quick temper, often provoked by trivialities, and his tendency, on occasion, to be coldly rude to those he regarded as misguided. Nor, when his temper seized him, was he discriminating on whom his wrath should fall. The originality of his work inevitably resulted in conflicts with accepted authority, and this led to many controversies, which he entered into with vigour. He could be unforgivingly hostile to those who in his opinion criticized his work unjustly, particularly if he suspected they were attempting to gain credit thereby. Nevertheless he undoubtedly enjoyed the cut and thrust of scientific controversy. His pungent verbal comments were well-known; though frequently made without malice, they were nevertheless disconcerting to those of less robust temperament. On scientific matters he was uncompromising, and was intolerant of scientific pretentiousness in all its forms, especially the pretentiousness of mathematicians.

As a child, Fisher sometimes showed an almost paranoiac reaction to the frustration of his will and to those he associated with such frustration. Naive and immature in personality, he seemed to have skipped the usual emotional development of adolescence. He grew up without developing a sensitivity to the ordinary humanity of his fellows and was unaware of the effects of his own behavior on others. His friends found him extremely controlling; they felt compelled to abandon their own wishes in his presence. Owing to his lack of empathy and, perhaps, an inability to interpret nonverbal behavior, he could be a poor judge of character. These are all characteristics of Asperger syndrome.

Fisher was difficult to know, to some natures baffling, to others intolerable. He was at once exceedingly self-centered and utterly self-forgetful. He spoke with clear diction and incisive, slow delivery. His conversation was not always easy to follow, but it was always fresh and stimulating. Fisher took immense pleasure in the process of thinking, the play of ideas, and the solution of puzzles; his logical and coherent argument was full of complexities, subtleties, and intriguing subplots. Among strangers he could be awkward and tongue-tied. When *savoir faire* was demanded, he might lose his head and have recourse to rudeness. At petty annoyances he was liable to go berserk. He was absent-minded and negligent of routine details.

Fisher married in 1917; he had eight children, two sons and six daughters. He was extremely critical of his wife, who was the victim of his volcanic rages, although she appeared only too anxious to support, help, and please her husband. To all appearances his accusations were wildly unfair and his anger a cruel and sadistic persecution. These harsh statements are taken from the plain-spoken biography by his daughter Joan Fisher Box (1978). We do not have Fisher's side of the story, but it is well established that marriage can be extraordinarily stressful for persons with Asperger syndrome and not uncommonly breaks down, as Fisher's did.

Fisher's highly intelligent but naive father came from a family in which intellectual brilliance was often accompanied by problems of personality. As one member of the family said, some Fishers were dull, some very sane and responsible, some were brilliant but went off the rails, some just went off the rails. Fisher's mother seemed cold and insensitive to outsiders, selfish, indolent, and arrogant to relatives. Their son's personality traits are likely to have come from her side of the family, but perhaps also from the father's side.

Norbert Wiener

The early years of Wiener's life are described in Chapter 2, so we begin this profile with his adolescence. The young Wiener, although socially inept, managed to get through the crises induced by the unnatural regime of his early years. By the age of 18 he had obtained a Ph.D. from Harvard, having graduated from Tufts College four years previously. His abilities were many-sided; he tried biology and philosophy before deciding to make his career as a mathematician. After an uncertain start he obtained a position at MIT, which was to be his base for the rest of his life. In the mid-1920s his younger brother, Fritz, was institutionalized with a diagnosis of schizophrenia. Wiener agonized over his brother's decline and worried constantly that he might suffer the same fate himself.

A successful marriage in 1926 helped him cope with many of the problems of everyday existence; he needed someone to take care of him and organize his life. His wife Margaret was a former student of his father's who had emigrated to America from Silesia at the age of 14. She made no secret of her admiration for Hitler, despite the fact that her husband was Jewish, as were many of their friends. They had two children, both daughters. When the girls were teenagers and began to acquire boyfriends, Margaret made their lives miserable by accusing them of nonexistent sexual delinquencies. As a result of her paranoid accusations, both daughters left home as soon as they could and thereafter had little contact with their parents.

Inquisitive, gregarious, and garrulous, Wiener made a habit of walking around MIT, waving his ever-present cigar, expounding his latest thoughts in his booming voice, his head thrown back like that of a trumpeting elephant, often tossing peanuts in the air and catching them in his mouth. He loved swimming and could often be found in the MIT swimming pool, still holding forth. He was the butt of all sorts of stories making fun of his absent-mindedness, his sense of insecurity, his lack of manual dexterity and, especially, his poor eyesight. Because he was afraid that he might become blind, he would practice finding his way around the corridors of MIT entirely by touch, and if he came to a classroom where a lecture was in progress would feel his way all round the walls, to the amazement of the teacher and students in the room.

Wiener displayed all the signs of cyclothymia, as Conway and Siegelman (2005) have suggested. He experienced mood swings, from persistent states of extreme elation to persistent states of sadness. Other textbook symptoms include "people-seeking," "pressure to keep talking," "thoughts of suicide," "feelings of worthlessness," and "volcanic rages," all of which he displayed, and yet through all this he remained surprisingly productive. Wiener was one of the outstanding American mathematicians of his generation, and the press hailed him as the American Leibniz after his death in 1964 at the age of 69.

In 1953 Wiener published the first installment of his autobiography, from which we have quoted above. As Hans Freudenthal remarked: "Although it conveys an extremely egocentric view of the world, I find it an agreeable story and not offensive, because it is naturally frank and there is no pose, least of all that of false modesty. All in all it is abundantly clear that he never had the slightest idea of how he appeared in the eyes of others" (James 2002). More objective accounts of Wiener's life have been given by Levinson (1966) and most recently by Conway and Siegelman (2005). According to Dirk Struik, who knew Wiener well:

> He was a man of enormous scientific vitality which the years did not seem to diminish, but this was complemented by extreme sensitivity. Wiener was a man of many moods, and these were reflected in his lectures, which ranged from among the worst to the very best I have ever heard. Sometimes he would lull his audience to sleep or get lost in his own computations—but on other occasions I have seen him hold a group of colleagues in breathless attention while he set forth his ideas in truly brilliant fashion. (Rowe 1989)

To quote Freudenthal again: "In appearance and behaviour, Norbert Wiener was a baroque figure, short, rotund and myopic, combining these and many other qualities in an extreme degree. His conversation was a curious mixture of pomposity and wantonness. He was a poor listener. His self-praise was playful, convincing, and never offensive. He spoke many languages but was not easy to understand in any of them."

Paul Dirac

While still a young man, Paul Dirac created an original and powerful formulation of the theory of quantum mechanics. He also created a primitive but important version of quantum electrodynamics, the relativistic wave equation of the electron, a theory of magnetic monopoles, and the notion of antiparticles. Sadly, few of his later contributions were of comparable importance, and none has the revolutionary character of his earlier work. Niels Bohr described Dirac as one of the century's great contributors, always going his own way, not making a school, compelled by the need for beauty and simplicity in physical theory, in later years more addicted to mathematics than was good for his physics.

The future physicist was born on 8 August 1902 in a suburb of Bristol. His mother Florence (née Hilten), the daughter of a ship's captain, came from Cornwall; she had worked in the library of Bristol University before marriage. His father, Charles Dirac, was brought up in the French-speaking Valais region of Switzerland but left home and emigrated to England in the 1880s. A gifted linguist and a strict disciplinarian, he made a career as a French schoolteacher. The Diracs had three children: Reginald, who was two years older than Paul, and Beatrice, four years younger than Paul. All three were registered at birth as Swiss citizens, but they and their father relinquished Swiss citizenship and became British in 1919. It has often been suggested that the strange patterns of behavior noticed in Paul were inherited from his father.

Charles wished his children to speak French at home, and speak it correctly. "My father made the rule that I should only speak to him in French," Paul recalled. "He thought it would be good for me to learn French in that way. However, since I found I couldn't express myself in French, it was better for me to stay silent than talk in English. So I became very silent at that time—that started very early" (James 2003b). Since his mother could hardly speak French at all, she usually took meals separately with the other children.

Paul Dirac's mathematical ability became apparent at an early age. The school where his father taught was particularly strong in mathematics and science, so the boy went there when he was 12. The academic standards were high, but the teaching had a vocational orientation. Modern languages were taught for use, metalwork was in the syllabus, as was shorthand, but classics and literature were not. In mathematics Dirac was soon so far ahead of his class that he was allowed to work largely on his own. He was recognized as a boy of exceptional intelligence.

After he grew up, Dirac's relations with his father remained frigid, and they had little personal communication, although Charles was proud of his son's success and tried to understand his work. When Dirac was awarded the Nobel Prize in 1933 and was told that he could invite his parents to attend the ceremony in Stockholm, he chose to invite only his mother.

After leaving school in 1918, Dirac entered Bristol University, where he studied electrical engineering. In 1921, after graduating, he looked without success for work as an engineer. (His elder brother Reginald had also been made to study engineering, although he wanted to be a doctor, but he disliked his first job and after six months committed suicide, following trouble with his girlfriend.) Paul won an exhibition to St. John's College, Cambridge, but did not take it up because of the cost of being a Cambridge undergraduate. Instead he took up the offer of two years' free tuition at Bristol University, where, unlike Cambridge, he would be able to economize by living at home. After a brilliant performance in the final examination, he was awarded a government grant that enabled him to go to Cambridge as a graduate student. There his research supervisor, Ralph Fowler, recognized in Dirac a student of exceptional ability. Of these years Dirac later recalled, "I confined myself entirely to scientific work, and continued at it pretty well day after day, except on Sundays when I relaxed and, if the weather was fine, I took a long solitary walk out in the country" (James 2003b). Before long Dirac was publishing research, at first on statistical mechanics, later on quantum mechanics. He always worked on fundamental problems, never minor ones. In Cambridge he rarely discussed physics with anyone else.

Dirac's first paper on quantum mechanics, the basis for his Ph.D. thesis of 1926, paralleled much of what was being done at the same time elsewhere, but he followed it up with more innovative work that immediately attracted the attention of theoreticians everywhere. Fowler arranged for Dirac to spend some time first in Copenhagen and then in Göttingen and Leiden. Dirac enjoyed the informal and friendly atmosphere at Copenhagen: "I admired Bohr very much. We had long talks together, very long talks in which Bohr did practically all of the talking" (James 2003b). While respecting Bohr greatly for his depth of thought, he said he did not know that Bohr had any influence on his own work because Bohr tended to argue qualitatively, while Dirac was more mathematical. At Leiden there was Paul Ehrenfest, while at Göttingen there were Max Born and his research student Robert Oppenheimer. On one occasion Oppenheimer offered Dirac some books to read, but Dirac politely refused, saying that reading books interfered with thought.

In 1927, Dirac was elected fellow of his Cambridge college; until his marriage he lived and worked there. Three years later he was elected to the Royal Society, although only 28, and then made the first of a number of visits to the Institute for Advanced Study in Princeton. In 1932 Cambridge elected him Lucasian Professor of Mathematics, his teacher Fowler having been elected Plumerian Professor of Mathematical Physics the previous year. Although not much interested in teaching, Dirac had some outstanding research students, who survived his sink-or-swim methods. As for his lectures, one who attended his regular course on quantum theory recalled: "The delivery was always exceptionally clear and one was carried along in the unfolding of an argument which seemed as majestic and inevitable as the development of a Bach fugue" (James 2003b).

In 1933 Dirac and Schrödinger shared the Nobel Prize for physics, "for the discovery of new productive forms of atomic theory." At first Dirac was inclined to refuse it because he so disliked publicity, but he changed his mind when he was warned that a refusal would attract even greater publicity. The following year Dirac returned to Princeton, where he developed a close friendship with Eugene Wigner, a professor of physics at the university and one of several outstanding mathematicians and physicists from Hungary who had emigrated to the United States. Wigner's widowed sister Margit Balsas was visiting from Budapest at the time. Her temperament was quite unlike Dirac's, spontaneous and impulsive with strong likes and dislikes. Margit introduced him to classical music; he listened to it on the radio but never went to concerts because he found the audience coughing too distracting. Theirs was an attraction of opposites, and in 1937 they were married in London. He is reputed to have introduced Margit not as his wife but as Wigner's sister at first. Their Cambridge household eventually included a son and a daughter from Margit's first marriage, both of whom took the name of Dirac, and two daughters from the second marriage; Dirac's mother Florence came to live with them after his father died. Margit Dirac described Paul as "too aloof" with their children, but the marriage seems to have worked, and he became more sociable in later years.

Until he retired in 1969, Dirac chose to remain in Cambridge, so as to encourage young British theoreticians, although he made extended visits to the Institute for Advanced Study in Princeton, the Tata Institute for Fundamental Research in Bombay, and Moscow State University. His wife never really cared for Cambridge, where many people were unfriendly to her, so in 1972 the Diracs emigrated to the United States and settled down at the Tallahassee campus of Florida State University. In Tallahassee Dirac was on the campus all day; whereas in Cambridge he always worked at home except when he gave classes and seminars. He published prolifically, over sixty papers during the last twelve years of his life, although these were not research papers. However, his powers were waning and his health was declining. He died at Tallahassee on 20 October 1984, at the age of 82.

At the funeral one of those present described his character in the following words: "In most social interactions he was mild-mannered, gentle, soft-spoken, reticent, modest, concise, unobtrusive, reserved, and unpretentious. Personally he was self-disciplined, strong-willed, resolute, firm, self-reliant, independent, persevering, stubborn, and tireless. In intellectual work he was meticulous mathematically and scientifically exact, rigorous, concise, honest, clear-thinking, courageous, self-sufficient, and tenacious" (James 2003b). Among the few honors he was prepared to accept, apart from the Nobel Prize, were the Copley and Royal medals of the Royal Society and the Order of Merit.

There are many stories about Dirac's personality, usually related to his taciturnity, which he was inclined to blame on his father, perhaps unfairly. A colleague in Cambridge, who had known him for years, said "I still find it very difficult to talk with Dirac. If I need his advice I try to formulate my question as briefly as possible." The response would come as from the witness stand. "He looks for five minutes at the ceiling, five minutes at the windows, and then says 'yes' or 'no.' And he was always right" (James 2003b). Dirac responded factually to direct questions, and the five-word answer might take five days to comprehend. He told Bohr, who was as voluble as Dirac was silent, that when he was young he learned that he should not start a sentence unless he knew how to finish it—not a recipe for spontaneous conversation. Once at question time after a lecture, someone stood up and said "I do not understand the derivation . . ." Dirac made no response, on the grounds that this was a statement, not a question. This is typical of people with Asperger syndrome, who tend to take everything literally.

In later years Dirac became an inveterate traveler, although he rarely visited Switzerland because he associated it so much with his father. He was often in the United States and often in the Soviet Union. When a research student who was thinking of going with him to Russia told him she was not sure she liked caviar, he replied that there was not much point in going to Russia if she did not.

As a young man Dirac never had a girlfriend, and seems to have had a rather Platonic conception of the opposite sex for some time. He confided to a friend: "I never saw a woman naked, either in childhood or in youth . . . The first time I saw a woman naked was in 1927, when I went to Russia with Peter Kapitza. She was a child, an adolescent. I was taken to a girls' swimming pool, and they bathed without swimming suits. I thought they looked nice." Heisenberg, who knew Dirac well, recalled:

We were on the steamer from America to Japan and I liked to take part in the
social life on the steamer and so, for instance, I took part in the dances in the
evening. Paul, somehow, didn't like that too much but he would sit in a chair
and look at the dances. Once I came back from a dance and took a chair beside
him and he asked me "Heisenberg, why do you dance?" I said "Well, when
there are nice girls it is a pleasure to dance." He thought for a long time about it
and after about five minutes he said "Heisenberg, how do you know be-
forehand that the girls are nice?" (James 2003b)

A German scientist wrote after his death:

Dirac was tall, gaunt, awkward and extremely taciturn. He has succeeded in
throwing everything he has into one dominant interest. He was a man, then, of
towering magnitude in one field, but with little interest and competence left for
other human activities. In other words he was the prototype of the superior
mathematical mind; but while in others this has coexisted with a multitude of
interests, in Dirac's case everything went into the performance of his great historical mission, the establishment of the new science quantum mechanics, to
which he probably contributed as much as any other man. (James 2003b)

As of yet, there exists no full-scale biography of Dirac. The main sources
for this profile are Dalitz and Peierls (1986), Goddard (1998), and Kragh
(1990). There is little doubt that Dirac met the criteria for Asperger
syndrome.

Kurt Gödel

Kurt Gödel was born on 28 April 1906 in the Moravian city of Brno, then in
Austro-Hungary but now part of the Czech Republic. His father Rudolf,
whose family was Sudeten German, was said to be of a ponderous and
serious disposition, although practical and energetic. Rudolf, a strict Lutheran, was director of and partner in one of the leading textile firms in the
city. His wife Marianne, the dominant parent, was of Rhenish stock; she was
a capable and imaginative mother to her two sons. Her younger son, the
timid and touchy Kurt, looked back on his childhood as a generally happy
time. However, at the age of 8 he had a severe bout of rheumatic fever and,
although he resumed normal life afterward, he managed to convince himself quite wrongly that he had a weak heart. Later in life he tended to seek
medical help when it was not necessary, but did not seek it when it was. He
hated to be touched. He gave up swimming and calisthenics, activities he

used to enjoy, preferring to stay at home and read. He developed an interest in languages that continued beyond his school days. Another early interest was theoretical physics; later he invented a version of the theory of relativity.

According to his biographer John Dawson (1997), Gödel was an exceptionally inquisitive child. He once asked an elderly visitor why her nose was so long. From the biography, the picture that emerges is of an earnest, serious, bright, sensitive child who was often withdrawn or preoccupied, and who from an early age exhibited signs of emotional instability.

Gödel's powers of concentrated study and sustained interest were already evident at school. If anyone asked him about a mathematical problem, he would start by writing it down in symbols before giving his response. He spoke slowly and calmly; his mind seemed very clear. His brother recalled that he had an individual and fixed opinion about everything, and could hardly be convinced to consider an alternative view. Someone who knew him for most of his life described him as a precocious youth who became old before his time.

In 1924 Gödel began studying at the University of Vienna. At first physics was his main field, but after a while he moved over to mathematics, in which the university was particularly strong at that time. His principal teacher was the analyst Hans Hahn, who was actively interested in the foundations of the subject and was a member of the Wiener Kreis, a band of positivist philosophers. Gödel attended their meetings regularly. Carl Menger, another participant, said that he never knew Gödel to take the floor at these meetings: he indicated interest solely by slight motions of the head. After graduation he wrote an impressive *habilitationsschrift* and became a *privatdozent* in 1933.

When Gödel's father died in 1929, his widowed mother moved from Brno to Vienna to be with her sons. Rudolf, the elder, had become a radiologist, wholly absorbed in his profession. Like his mother, Kurt sampled the rich cultural life of the Austrian capital. He became very attached to a nightclub dancer named Adele Porkert, who was a divorcée. She was described as garrulous, uncultured, sharp-tongued, and strong-willed. Despite his mother's disapproval, they quietly got married in 1938 after ten years of close friendship.

Gödel became more and more withdrawn and emaciated as he got older. He ignored a bleeding stomach ulcer until it was nearly too late. He had a fear of poisoning, for example, by gases that might be escaping from the refrigerator in his kitchen or from the paint on the radiator in his room.

Adele was expected to taste his food before he ate it. He kept a record of his body temperature and ingested a variety of pills for imaginary heart problems. Shortly after publication of his famous work on the consistency of the continuum hypothesis, he exhibited signs of depression so serious that his family, fearing that he might become suicidal like his father, caused him to be committed to a sanatorium. This was the first of several such mental breakdowns. He suffered psychotic episodes in 1954 and 1970.

Gödel, by and large, had the political views that were standard in the Austria of his youth. He was not Jewish, and seemed to view the rise of the Nazis with equanimity. When Austria became part of Germany in 1938, he left for the United States to avoid being conscripted into the German army. He spent his first term in the United States in Princeton at the Institute for Advanced Study, a second at the University of Notre Dame, and then returned to the Institute in Princeton, where he was based for the rest of his life. He was not given a permanent appointment until 1953, however, because of concern about his health and mental stability. When certain foreign visitors came to Princeton, he would not leave home, because he feared that they might try to kill him. Other well-known mathematician visitors he simply declined to recognize.

The Gödels were not very sociable. Adele was liable to form quite the wrong impression of other people; she would take mortal offense at some imagined slight. Gödel would only open up in the company of a few fellow mathematicians, mostly European émigrés like himself. Among his colleagues his only close friend was Einstein, who remarked that he went to the Institute for Advanced Study every day "just to have the privilege of walking home with Kurt Gödel." He liked the United States and became an American citizen in 1948. The examining judge made the unfortunate error of asking him his opinion of the Constitution of the United States, unleashing a pent-up lecture on its inconsistencies.

After Gödel developed a prostate condition around 1970, he became depressed and withdrew more and more from the world around him. Although his mind remained clear until the end, his dread of poisoning led to semistarvation, and he died in the Princeton hospital on 14 January 1978, at the age of 71. Dawson (1997) sums up his biography of Gödel by describing him as a reclusive genius whose work has generally been considered abstruse and whose life, combining elements of rationality and psychopathology, has been the subject more of rumor than of factual knowledge.

Gödel was a private person, sensitive and hypochondriac, lacking warmth and sensibility. He avoided people who made him feel uncomfortable and did not care to contribute to nonmathematical conversations with strangers. He was extremely solemn, very serious, quite solitary, and distrustful of common sense as a means of arriving at the truth. Despite his prodigious intellect, he often exhibited a childlike naivety. His favorite film was Walt Disney's *Snow White and the Seven Dwarfs*. Gödel's tastes remained unsophisticated and his well-being dependent on the efforts of those who were willing to shield him from the outside world and ensure he was looked after, especially in times of physical or mental stress.

Gödel was interested in comparative religion, as attested by the many books in his library devoted to strange religious sects. Sometimes he was so obsessed by his work that, like Newton, he would forget to eat his meals. He retained almost every scrap of paper that ever crossed his desk: library request slips, luggage tags, cranky correspondence, letters from autograph seekers. However warm the weather, he would dress up in sweaters, overcoat, and ear muffs for his daily stroll with Einstein, his only close friend, to the nearby Institute for Advanced Study. They didn't want to speak to anyone else, explained a colleague; they only wanted to speak to each other.

Philosophically Gödel was a Platonist. For him the world was rational; there were no fortuitous events. An arch-rationalist like Gödel is determined to find hidden causes to account for human behavior, making him distrustful of human motives. He eschewed controversy and held back from open criticism of others. It was his habit to listen to what others had to say, only occasionally interjecting incisive comments. He published very sparingly; much of his work appeared posthumously. Fitzgerald and Lyons (2004) argue that Gödel had Asperger syndrome.

References

Note on Sources: The main sources for the profiles are listed below. Direct quotes in the profiles are taken from these sources, unless otherwise noted. Reference books, especially the *Dictionary of Scientific Biography* (1970–1980), have also been consulted.

Lagrange: Sarton (1944)
Gauss: Dunnington (1960), Hall (1970), Kaufmann-Bühler (1981)
Cauchy: Belhoste (1991)
Dirichlet: Butzer (1987)
Hamilton: Hankins (1980), O'Donnell (1983)
Galois: Rigatelli (1996)
Byron: Moore (1977), Stein (1985), Toole (1992), Woolley (1999)
Riemann: Laugwitz (1999)
Cantor: Dauben (1979), Grattan-Guinness (1971)
Kovalevskaya: Kennedy (1983), Koblitz (1983, 1987)
Poincaré: Toulouse (1910)
Hilbert: Courant (1981), Reid (1970)

Hadamard: Maz'ya and Shaposhnikova (1998)
Hardy: Hardy (1940), Kanigel (1991)
Noether: Brewer and Smith (1981), Dick (1981)
Ramanujan: Berndt and Rankin (1995), Bollobas (1998), Fitzgerald (2004), Kanigel (1991), Ranganathan (1967)
Fisher: Box (1978), Yates and Mather (1963)
Wiener: Conway and Siegelman (2005), Masani (1990), Levinson (1966), Wiener (1953, 1956)
Dirac: Dalitz and Peierls (1986), Goddard (1998), Kragh (1990)
Gödel: Dawson (1997), Goldstein (2005), Kreisel (1980)

Albers, D. J. and Alexanderson, G. L. (1985) *Mathematical People: Profiles and Interviews.* Boston, MA: Birkhauser.
Albers, D. J., Alexanderson, G. L., and Reid, C. (1997) *More Mathematical People: Contemporary Conversations.* San Diego, CA: Academic Press.

Alexanderson, G. L. (1999) *The Random Walks of George Pólya*. Washington, DC: Mathematical Association of America.

Alexandrov, P. S. (1979/1980) Pages from an autobiography. (trans. A. Dowker) *Russian Mathematical Surveys* I: 34 (1979), 267–304; II: 35 (1980), 315–58.

Anastasi, A. (1958) *Differential Psychology*. New York: Macmillan.

Anderson, N., O'Connor, N., and Hermelin, B. (1998) A specific calculating ability. *Intelligence* 26: 383–403.

Andrews, G. E., et al., eds. (2000) *Kolmogorov in Perspective*. (trans. H. H. McFaden) Providence, RI: American Mathematical Society and London Mathematical Society.

Archibald, R. C. (1924) Mathematics and music. *American Mathematical Monthly* 31: 1–25.

Arnold, V. I. (1993) On A. N. Kolmogorov (trans. E. Primrose). In S. Zdravkovska and P. L. Duren, eds., In Golden Years of Moscow Mathematics, *History of Mathematics*, vol. 6. Providence, RI: American Mathematical Society.

Asperger, H. (1944) Die "autistischen psychopathen" im kindesalter, *Archives für Psychiatrie und Nervenkrankheiten* 117: 76–136. Translated in U. Frith, ed., *Autism and Asperger's Syndrome*. Cambridge: Cambridge University Press (1991), 37–92.

Barlow, G. (1952) *Mental Prodigies*. New York: Greenwood Press.

Baron-Cohen, S. (2003) *The Essential Difference: Men, Women, and the Extreme Male Brain*. London: Allen Lane.

Baron-Cohen, S., Bolton, P., Wheelwright, S., Scahill, V., Short, L., Mead, G., and Smith, A. (1998) Does autism occur more often in families of physicists, engineers, and mathematicians? *Autism* 2: 296–301.

Baron-Cohen, S., Richler, J., Bisarya, D., Gurunathan, N. and Wheelwright, S. (2003) The systemizing quotient: an investigation of adults with Asperger's syndrome or high-functioning autism, and normal sex differences. *Philosophical Transactions of the Royal Society, Series B* (special issue on autism, mind and brain) 358: 361–740.

Baron-Cohen, S. and Wheelwright, S. (2001) The autism-spectrum quotient (AQ): Evidence from Asperger's syndrome/high-functioning autism, males and females, scientists and mathematicians. *Journal of Autism and Developmental Disorders* 31: 5–17.

Baron-Cohen, S., Wheelwright, S., Stone, V., and Rutherford, M. (1999) A mathematician, a physicist, and a computer scientist with Asperger's syndrome: Performance on folk psychology and folk physics test. *Neurocase* 5: 475–83.

Baron-Cohen, S., Wheelwright, S., Burdenshaw, A., Hobson, E. (forthcoming) Mathematical talent is linked to autism. *Human Nature*.

Baum, J. (1986) *The Calculating Passion of Ada Byron*. Hamden, CT: Shoe String Press.

Belhoste, B. (1991) *Augustin-Louis Cauchy* (trans. F. Ragland). Berlin: Springer Verlag.

Bell, E. T. (1937) *Men of Mathematics*. London: Victor Gollanz.

Benbow, C. P. (1987) Possible biological correlates of precocious mathematical reasoning ability. *Trends in the Neurosciences* 10: 17–20.

———. (1988a) *Neuropsychological Perspectives on Mathematical Talent in the Exceptional Brain*. New York: Guilford Press.

———. (1988b) Sex differences in mathematical reasoning ability in intellectually talented preadolescents: Their various effects and possible causes. *Behavioural and Brain Sciences* 11: 169–232.

Benbow, C. P. and Lubinski, D. (1992) *Psychological Profiles of the Mathematically Talented: Some Sex Differences and Evidence Supporting Their Biological Basis in Origin and Development of High Ability*. CIBA Foundation Symposium. New York: Wiley.

Benbow, C. P. and Stanley, J. C. (1983) Sex differences in mathematical reasoning ability: More facts. *Science* 222: 1931–2029.

Berndt, B. C. (2001) *Ramanujan: Essays and Surveys*. Providence, RI: American Mathematical Society and London Mathematical Society.

Berndt, B. C. and Rankin, R. A. (1995) *Ramanujan: Letters and Commentary*. Providence, RI: American Mathematical Society.

Binet, A. (1894) *Psychologie des grands calculateurs et joueurs d'échecs*. Paris: Hachette.

Bollobas, B. (1988) Ramanujan: A glimpse of his life and his mathematics. *Cambridge Review*, 76–80.

Box, J. F. (1978) *R.A. Fisher, The Life of a Scientist*. Chichester, UK: Wiley.

Brain, R. (1960) *Some Reflections on Genius*. New York: Pitman.

Brewer, J. W. and Smith, M. K. (1981) *Emmy Noether: A Tribute to Her Life and Work*. New York: Marcel Dekker.

Bunyan, W. M. (1984) *Life Histories and Psychobiography: Explorations in Theory and Method*. Oxford: Oxford University Press.

Butterworth, B. (1999) *The Mathematical Brain*. London: Macmillan.

———. (2005) The development of arithmetical abilities. *Journal of Child Psychology and Psychiatry* 46: 3–18.

Butzer, Paul L. (1987) Dirichlet and his role in the founding of mathematical physics. *Archives internationales d'histoire des sciences* 37: 49–82.

Campbell, J. I. D., ed. (2005) *Handbook of Mathematical Cognition*. Hove, UK: Psychology Press.

Cartan, E. (1952–1955). *Oeuvres complètes*. Paris: Gauthiers-Villars.

Case, B. A. and Leggett, A. M., eds. (2005) *Complexities: Women in Mathematics*. Princeton, NJ: Princeton University Press.

Changeux, J.-P. and Connes, A. (1995) *Conversations on Mind, Matter and Mathematics* (ed. and trans. M. B. DeBevoise) Princeton, NJ: Princeton University Press.

Conway, F. and Siegelman, J. (2005) *Dark Hero of the Information Age: In Search of Norbert Wiener—The Father of Cybernetics*. New York: Basic Books.

Courant, R. (1981) Reminiscences of Hilbert's Göttingen (ed. John Ewing). *Mathematical Intelligencer* 3: 154–64.

Courant, R. and Robbins, H. (1941) *What Is Mathematics?* New York: Oxford University Press.

Cox, C. M. (1926) *The Early Mental Traits of Three Hundred Geniuses.* Stanford, CA: Stanford University Press.

Dalitz, R. H. and Peierls, R. (1986) Paul Adrien Maurice Dirac. *Biographical Memoirs of Fellows of the Royal Society* 32: 137–85.

Darboux, G. (1916/1956) Eloge historique d'Henri Poincaré. In P. Appell, ed., *Oeuvres d'Henri Poincaré,* vol II. Paris: Gauthiers Villars.

Dauben, J. W. (1979) *Georg Cantor: His Mathematics and Philosophy of the Infinite.* Princeton, NJ: Princeton University Press.

Davis, P. J. and Hersh, R. (1990) *The Mathematical Experience.* London: Penguin.

Dawson, J. W. (1997) *The Life and Work of Kurt Gödel.* Wellesley, MA: A. K. Peters.

Dedekind, R. (1930) *Gesammelte Werke* (ed. Fricke, R., et al). Brunswick: Vieweg.

Dehaene, S. (1997) *The Number Sense: How the Mind Creates Mathematics.* Oxford: Oxford University Press.

Dehn, M. (1983) The mentality of the mathematician (trans. A. Schenitzer). *Mathematical Intelligencer* 5(2): 18–28.

Denis, M., Mellet, E., and Kosslyn, S., eds. *Neuropsychology of Mental Imagery.* Hove, UK: Psychology Press.

Devlin, K. (2000) *The Maths Gene: Why Everyone Has It but Most People Don't Use It.* London: Weidenfeld and Nicholson.

Diamond, A. M. (1986) The life-cycle research productivity of mathematicians and scientists. *Journal of Gerontology* 41: 520–25.

Dick, A. (1981) *Emmy Noether, 1882–1935.* Basle: Birkhäuser Verlag.

Dinnage, R. (2004) *Alone! Alone! Lives of Some Outsider Women.* New York: New York Review of Books.

Dunnington, G. W. (1960) *Carl Friedrich Gauss.* New York: Stecher-Hafner.

Einstein, A. (1951) Autobiographical notes (trans. P. A. Schilpp). In P. A. Schilpp, ed., *Albert Einstein: Philosopher Scientist.* New York: Tudor.

Eisenstadt, J. M. (1978) Parental loss and genius. *American Psychologist* 33: 211–33.

Ellis, H. (1904) *A Study of British Genius.* London: Hurst and Blackett.

Emmer, M., ed. (1993) *The Visual Mind: Art and Mathematics.* Cambridge, MA: MIT Press.

Evans, J. (2000) *Adults' Mathematical Thinking and Emotions. A Study of Numerate Practices.* London: Routledge Falmer.

Fauvel, J., Flood, R., and Wilson, R. (2003) *Music and Mathematics: From Pythagoras to Fractals.* Oxford: Oxford University Press.

Feldman, D. H. (1986) *Nature's Gambit.* New York: Basic Books.

Fias, W., and Fischer, M. H. (2005) Spatial recognition of numbers. In J. I. D. Campbell, ed., *Handbook of Mathematical Cognition.* Hove, UK: Psychology Press.

Field, J. V. (2005) *Piero della Francesca: A Mathematician's Art*. New Haven, CT: Yale University Press.

Fitzgerald, M. (2000) Is the cognitive style of persons with Asperger's syndrome also a mathematical style? *Journal of Autism and Developmental Disorders* 30: 175–76.

———. (2001) Did Lord Byron have attention deficit hyperactivity disorder? *Journal of Medical Biography* 9: 31–33.

———. (2002a) Did Ramanujan have Asperger's disorder or Asperger's syndrome? *Journal of Medical Biography* 10: 1–3.

———. (2002b) Asperger's disorder and mathematicians of genius. *Journal of Autism and Developmental Disorders* 32: 59–60.

———. (2004) *Autism and Creativity: Is There a Link between Autism in Men and Exceptional Ability?* Hove, UK: Brunner-Routledge.

———. (2005) *The Genesis of Artistic Creativity: Asperger's Syndrome and the Arts*. London: Jessica Kingsley.

Fitzgerald, M., and Lyons, V. (2004) Kurt Gödel (1906–1978): The mathematical genius who had Asperger syndrome. *Autism/Asperger's Digest*.

Fowler, W. (1983) *Potentials of Childhood*. Lexington, MA: Lexington Books.

Frith, U., ed. (1991) *Autism and Asperger Syndrome*. Cambridge: Cambridge University Press.

Frith, U. (2003) *Autism: Explaining the Enigma*. Oxford: Blackwell.

Fuegi, J. and Francis, J. (2003) Lovelace and Babbage and the creation of the 1843 notes. *Annals of the History of Computing*, 16–25.

Galton, F. (1869) *Hereditary Genius, an Enquiry into its Laws and Consequences*. London: Macmillan.

———. (1874/1974) *English Men of Science: Their Nature and Nurture*. London: Macmillan.

———. (1883/1951) *Inquiries into Human Faculty and Its Development*. London: Dent.

Gardner, H. (1997) *Extraordinary Minds*. New York: Basic Books.

Gardner, H. J. and Wilson, R. J. (1993) Thomas Archer Hirst: Mathematician Xtravagant III: Göttingen and Berlin. *American Mathematical Monthly* 100: 619–25.

Geary, D. (1996) Sexual selection and sex differences in mathematical studies. *Behavioral Brain Science* 19: 229–84.

Gillberg, C. (2002) *A Guide to Asperger Syndrome*. Cambridge: Cambridge University Press.

Gillberg, C. and Coleman, M. (2000) *The Biological Basis of Autism*. Cambridge: MacKeith Press.

Goddard, P., ed. (1998) *Paul Dirac: The Man and His Work*. Cambridge: Cambridge University Press.

Goertzel, M. G., Goertzel, V., and Goertzel, T. G. (1962/1978) *Three Hundred Eminent Personalities*. San Francisco, CA: Jossey-Bass.

Goertzel, V. and Goertzel, M. G. (1962) *Cradles of Eminence*. Boston, MA: Little, Brown.

Goldstein, R. (2005) *Incompleteness: the Proof and Paradox of Kurt Gödel*. New York: W. W. Norton.

Grandin, T. (1992) An inside view of autism. In Schopler, E., and Mesibov, G. B., eds., *High-Functioning Individuals with Autism*. New York: Plenum Press.

———. (1996) *Thinking in Pictures*. New York: Vintage.

Grattan-Guinness, I. (1971) Towards a biography of Georg Cantor. *Annals of Science* 27: 345–91.

Graves, R. P. (1889) Life of Sir William Rowan Hamilton (3 vols.). Dublin: Hodges and Figgis.

Gregory, R. L. (2005) *The Oxford Companion to the Mind*. Oxford: Oxford University Press.

Hadamard, J. (1945) *The Psychology of Invention in the Mathematical Field*. Princeton, NJ: Princeton University Press.

Hall, T. (1970) *Carl Friedrich Gauss*. Cambridge, MA: MIT Press.

Halmos, P. (1985) *I Want to Be a Mathematician*. New York: Springer Verlag.

Halsted, G. B. (1946) *The Foundations of Science*. Philadelphia: Science Press.

Hankins, T. L. (1980) *Sir William Rowan Hamilton*. Baltimore: The Johns Hopkins University Press.

Happe, F. (1999) Autism: cognitive deficit or cognitive style? *Trends in Cognitive Sciences* 3(6): 216–22.

Hardy, G. H. (1921) Srinivasa Ramanujan. *Proceedings of the London Mathematical Society* 2(10): 40–58.

———. (1940/1967) *A Mathematician's Apology*. Cambridge: Cambridge University Press.

———. (1966–1979). *The Collected Papers of G. H. Hardy*. Oxford: Clarendon Press.

Helmholtz, H. von (1873) *Popular Lectures on Scientific Subjects*. London: Longmans, Green.

Helson, R. (1980) Creative mathematicians. In R. S. Albert, ed., *Genius and Eminence*. Oxford: Pergamon Press.

Henrion, C. (1997) *Women in Mathematics*. Bloomington, IN: Indiana University Press.

Hensel, S. (1881) *The Mendelssohn Family (1729–1847)* (trans. C. Klingemann). London: Sampson Low, Marston, Searl, and Rivington.

Hermelin, B. (2001) *Bright Splinters of the Mind*. London: Jessica Kingsley.

Hermelin, B. and O'Connor, N. (1990) Factors and primes: A specific numerical ability. *Psychological Medicine* 16: 885–93.

Hersh, R. (2001) Mathematical menopause, or, a young man's game? What is it like to be still a mathematician and no longer young? *Mathematical Intelligencer* 23(3): 52.

Hershman, D. J. and Lieb, J. (1998) *Manic Depression and Creativity*. Buffalo, NY: Prometheus.

Hines, M. (2004) *Brain Gender*. New York: Oxford University Press.

Hodges, A. (1983) *Alan Turing: The Enigma.* London: Burnett Books.

Holton, G. (1978) *Scientific Imagination. Case Studies.* Cambridge: Cambridge University Press.

Houston, R. and Frith, U. (2000) *Autism in History: The Case of Hugh Blair of Borgue.* Oxford: Blackwell.

Howe, M. J. A. (1989) *Fragments of Genius.* London: Routledge.

—— (1999) *Genius Explained.* Cambridge: Cambridge University Press.

Hunter, A. M. (1968) An exceptional talent for calculative thinking. *British Journal of Psychology* 53, 243–58.

——. (1977) An exceptional memory. *British Journal of Psychology* 68: 155–64.

Hyde, J. D., Fennema, E. and Lamon, S. J. (1990) Gender difference in mathematics performance: A meta-analysis. *Psychological Bulletin* 107: 139–55.

James, I. (2002) *Remarkable Mathematicians.* Cambridge: Cambridge University Press; Washington, DC: Mathematical Association of America.

——. (2003a) Autism in mathematicians. *Mathematical Intelligencer* 25: 62–65.

——. (2003b) *Remarkable Physicists.* Cambridge: Cambridge University Press.

——. (2003c) Singular scientists. *Journal of the Royal Society of Medicine* 96: 36–39.

——. (2005) *Asperger's Syndrome and High Achievement.* London: Jessica Kingsley.

Jamison, K. R. (1993a) *Touched with Fire.* New York: The Free Press.

——. (1993b) Mood disorders and patterns of creativity in British writers and artists. *Psychiatry* 52: 125–34.

Kac, M. (1985) *Enigmas of Chance.* New York: Harper and Row.

Kaluza, R. (1966) *Through a Reporter's Eyes: The Life of Stefan Banach* (trans. and ed. W. Woycynski and A. Kostant). Basle: Birkhäuser Verlag.

Kanigel, R. (1991) *The Man Who Knew Infinity: A Life of the Genius Ramanujan.* New York: Charles Scribner's Sons.

Karlsson, J. L. (1978) *Inheritance of Creative Intelligence.* Chicago, IL: Nelson Hall.

Kaufmann-Bühler, W. (1981) *Gauss: A Biographical Study.* Berlin: Springer Verlag.

Kelly, S. J., Macaruso, P., and Sokol, S. M. (1997) Mental calculation in an autistic savant: a case study. *Journal of Clinical and Experimental Neuropsychology* 19: 172–84.

Kennedy, Don H. (1983) *Little Sparrow: A Portrait of Sonya Kovalevskaya.* Athens, OH: Ohio University Press.

Keynes, Milo (1995) The personality of Isaac Newton. *Notes and Records of the Royal Society* 49: 1–56.

Keys, W., Harris, S., and Fernandes, C. (1996) *Third International Mathematics and Science Study.* Slough, UK: National Foundation for Educational Research.

Keyser, C. J. (1940) *The Human Worth of Rigorous Thinking.* New York: Scripta Mathematica.

Klein, C. F. (1926/1927) *Vorlesungen über die Entwicklung der Mathematik im 19 Jahrhundert.* Berlin: Julius Springer.

Kline, Morris (1953) *Mathematics in Western Culture.* New York: Oxford University Press.

——. (1980) *The Loss of Certainty*. New York: Oxford University Press.

Koblitz, A. (1983) *A Convergence of Lives*. Basel: Birkhäuser Verlag.

——. (1987) Sofia Kovalevskaia—a biographical sketch. In Linda Keen, ed., *The Legacy of Sonya Kovalevskaya*. Providence, RI: American Mathematical Society.

Kragh, H. (1990) *Dirac: A Scientific Biography*. Cambridge: Cambridge University Press.

Kreisel, G. (1980) Kurt Gödel. *Biographical Memoirs of Fellows of the Royal Society* 26: 149–224.

Kretschmer, E. (1931) *The Psychology of Men of Genius* (trans. T. B. Carrell). New York: Harcourt Brace.

Krutetskii, V. A. (1976) *The Psychology of Mathematical Abilities in School Children* (trans. J. Teller). Chicago, IL: University of Chicago Press.

Kûmmel, E. (1889–1897) Gedachtnisrede auf Gustav Peter Lejeune Dirichlet. In Kronecker, L., and Fuchs, L., eds. *G. Lejeune Dirichlet Werke*. Berlin: George Reiner.

Kursunoglu, B. N. and Wigner, E. P., eds. (1987) *Reminiscences about a Great Physicist: Paul Adrien Maurice Dirac*. Cambridge: Cambridge University Press.

Lakoff, G. and Nunez, R. (2000) *Where Does Mathematics Come From?* New York: Basic Books.

Lang, S. (1985) *The Beauty of Doing Mathematics*. New York: Springer.

Laugwitz, D. (1999) *Bernhard Riemann 1826–1866: Turning Points in the Conception of Mathematics*. Basle: Birkhäuser Verlag.

Ledgin, N. (2002) *Asperger's and Self-esteem: Insight and Hope through Role Models*. Arlington, TX: Future Horizons.

Leeson, D. N. (2000) Mozart and mathematics. *Mozart Jahrbuch 1999*. Kassel, Germany: Baerenreither.

Lehman, H. C. (1953) *Age and Achievement*. Princeton, NJ: Princeton University Press.

Levinson, N. (1966) Wiener's life. *Bulletin of the American Mathematical Society* 72(1), Pt II: 1–32.

Ludwig, A. M. (1992) Creative achievement and psychopathology: Comparisons among the professions. *American Journal of Psychotherapy* 46: 330–56.

Lykken, D. (1998) The genetics of genius. In A. Steptoe, ed., *Genius and the Mind*. Oxford: Oxford University Press.

Maccoby, E. E. and Jacklin, C. N. (1974) *The Psychology of Sex Differences*. Stanford, CA: Stanford University Press.

Masani, P. R. (1990) *Norbert Wiener 1894–1964*. Basle: Birkhäuser Verlag.

Maz'ya, V. and Shaposhnikova, T. (1998) *Jacques Hadamard: A Universal Mathematician*. History of Mathematics series, vol. 14. Providence, RI: American Mathematical Society.

McManus, I. C. (2002) *Right Hand, Left Hand*. London: Weidenfeld and Nicholson.

Miller, A. I. (1996) *Insights of Genius: Imagery and Creativity in Science and Art*. New York: Copernicus.

Minkowski, H. (1917) P. G. Lejeune Dirichlet und seine Bedeutung. In A. Speiser, H. Weyl, and D. Hilbert, eds., *Gesammelte Abhandlungen Hermann Minkowski*. Leipzig: B. G. Teubner.

Mitchell, F. D. (1907) Mathematical prodigies. *American Journal of Psychiatry* 18: 61–143.

Möbius, P. J. (1900) *Uber die Anlage zur Mathematik*, Leipzig: Johann Ambrosius Barth.

Monk, R. (1996) *Bertrand Russell: the Spirit of Solitude*. London: Vintage Books.

——. (1997) *Bertrand Russell: the Ghost of Madness*. London: Jonathan Cape.

Moore, D. L. (1977) *Ada, Countess of Lovelace: Byron's Legitimate Daughter*. London: John Murray.

Muir, A. T. (1988) The psychology of mathematical creativity. *Mathematical Intelligencer* 10: 33–37.

Nasar, S. (1998) *A Beautiful Mind: A Biography of John Forbes Nash, Jr.* New York: Simon and Schuster.

Nettle, D. (2006) *Strong Imagination: Madness, Creativity and Human Nature*. Oxford: Oxford University Press.

O'Donnell, S. (1983) *William Rowan Hamilton: Portrait of a Prodigy*. Dublin: Boole Press.

Ore, O. (1957) *Niels Henrik Abel*. Minneapolis, MN: University of Minnesota Press.

Osen, L. (1974) *Women in Mathematics*. Cambridge, MA: MIT Press.

Parikh, C. (1991) *The Unreal Life of Oscar Zariski*. San Diego, CA: Academic Press.

Perl, T. (1978) *Math Equals*. Reading, MA: Addison Wesley.

Pesenti, M. (2005) Calculation abilities in expert calculators. In J. Campbell, ed., *Handbook of Mathematical Cognition*. Hove, UK: Psychology Press.

Pesenti, M., Seron, X., and Samson, D. (1999) Basic and exceptional calculation abilities in a calculating prodigy; a case study. *Mathematical Cognition* 5: 97–148.

Poincaré, H. (1915–1956) *Oeuvres Henri Poincaré* (ed. P. Appell). Paris: Gauthier Villars.

——. (1945) Mathematical Invention. In Halsted, G. B., *The Foundations of Science*. Philadelphia: Science Press.

——. (1952) *Science and Method* (trans. F. Maitland). New York: Dover Publications.

Post, F. (1994) Creativity and psychopathology: a study of 291 world-famous men. *British Journal of Psychiatry* 165: 22–34.

Radford, J. (1990) *Child Prodigies and Exceptional Early Achievers*. Hemel Hempstead: Harvester Wheatsheaf.

Ramachandran, V. S. and Blakeslee, S. (1998) *Phantoms in the Brain*. New York: Morrow.

Ranganathan, S. R. (1967) *Ramanujan*. London: Asia Publishing House.

Reid, C. (1996) *Courant*. Berlin: Springer Verlag.

——. (1970) *Hilbert*. Berlin: Springer Verlag.

Révész, G. (1946) Die Beziehung zwischen mathematischer und musikalischer Begabung. *Schweizerische Zeitschrift für Psychologie und ihre Anwendungen* 5: 269–81.

Rigatelli, L. T. (1996) *Evariste Galois (1811–1832)*. Basle: Birkhäuser Verlag.

Roe, A. (1951) A study of imagery in research scientists. *Journal of Personality* 19: 459–70.

Rouse Ball, W. W. (1892) *Mathematical Recreations and Essays*. London: Macmillan.

Rowe, D. E. (1989) Interview with Dirk Jan Struik. *Mathematical Intelligencer* 11 (1).

Russell, B. (1903) *My Philosophical Development*. London: Routledge.

———. (1910) The study of mathematics. *Philosophical Essays*. London: Longmans, Green.

———. (1967/1968/1969) *The Autobiography of Bertrand Russell, 1872–1914, 1914–1944, 1944–1968*. Boston, MA: Atlantic Monthly Press.

Sacks, O. (1995) *An Anthropologist on Mars*. London: Picador.

Sarton, G. (1944) Lagrange's personality. *Proceedings of the American Philosophical Society* 88: 457–96.

Schilpp, P. A., ed. *Albert Einstein: Philosopher Scientist*. New York: Tudor.

Schoenfliess, A. (1922) Zur errinerung an Georg Cantor. *Jahresbericht der Deutchen Mathematiker Vereinigung* 31: 97–106.

Schoenfliess, A. (1927) Die Krisis in Cantor's Mathematischen Schaften. *Acta Mathematica* 50, 1–23.

Scripture, E. W. (1891) Arithmetical prodigies. *American Journal of Psychiatry* 4: 1–59.

Senechal, M. T. (2004) The mysterious Mr Ammann. *Mathematical Intelligencer* 26 (4): 10–21.

Seron, X., Pesenti, M., Noel, M. P., Deloche, G., and Cornet, J. A. (1992) Images of numbers, or "when 98 is upper left and 6 sky blue." *Cognition* 44: 159–96.

Simonton, D. K. (1994) *Greatness: Who Makes History and Why*. New York: Fulford Press.

Skaalvik, E. M. and Rankin, R. J. (1994) Gender differences in mathematics and verbal achievement, self-perception and motivation. *British Journal of Educational Psychology* 64: 419–428.

Skirbekk, V. (2003) *Age and Individual Productivity: A Literature Survey*. Rostock, Germany: Max Planck Institute for Demographic Research.

Smith, S. B. (1983) *The Great Mental Calculators: The Psychologies and Lives of Calculating Prodigies, Past and Present*. New York: Columbia University Press.

Snyder, A. W. and Mitchell, J. D. (1999) Is integer arithmetic fundamental to mental processing? The mind's secret arithmetic. *Proceedings of the Royal Society London*, Series B, 266: 597–592.

Ssucharewa, G. E. (1926) Die schizoiden psychopathien in kindesalter. *Monatschrift für Psychiatrie und Neurologie* 60: 235–61.

Stage, E. K., Kreinberg, N., Eccles, J., and Becker, J. R. (1985) Increasing the participation and achievement of girls and women in mathematics, science, and engineering. In S. S. Klein, ed., *Handbook for Achieving Sex Equity through Education.* Baltimore: The Johns Hopkins University Press.

Steen, L. A. (1978) *Mathematics Today.* New York: Springer Verlag.

Stein, D. (1985) *Ada: A Life and a Legacy.* Cambridge, MA: MIT Press.

Stern, N. (1978) Age and achievement in mathematics. *Social Studies in Science* 8: 127–40.

Storfer, M. D. (1990) *Intelligence and Giftedness.* San Francisco, CA: Jossey-Bass.

Storr, A. (1972) *The Dynamics of Creation.* New York: Charles Scribner's Sons.

Stubhaug, A. (2000) *Niels Henrik Abel and His Times.* Berlin: Springer Verlag.

——. (2002) *The Mathematician Sophus Lie.* Berlin: Springer Verlag.

Sulloway, F. J. (1996) *Born to Rebel: Birth Order, Family Dynamics, and Creative Lives.* London: Little, Brown.

Terman, I. M. (1925) *Mental and Physical Traits of a Thousand Gifted Children* Stanford CA: Stanford University Press.

Titchmarsh, E. C. (1949) Godfrey Harold Hardy. *Biographical Memoirs of Fellows of the Royal Society* 6: 447–61.

Toole, B. A. (1992) *The Enchantress of Numbers.* Mill Valley, CA: Strawberry Press.

Toulouse, E. (1910) *Henri Poincaré.* Paris: Flamarrion.

Treffert, D. A. (1988) The idiot savant: A review of the syndrome. *American Journal of Psychiatry* 145 (5): 563–72.

——. (1989) *Extraordinary People: Redefining the Idiot Savant.* New York: Harper and Row.

Trevor-Roper, P. (1988) *The World through Blunted Sight.* London: Allen Lane.

Ulam, S. M. (1976) *Adventures of a Mathematician.* New York: Charles Scribner's Sons.

Van Dalen, D. (1997) *Mystic, Geometer and Intuitionist: The Life of L. E. J. Brouwer.* Oxford: Clarendon Press.

Walker, A., and Fitzgerald, M. (2006) *Unstoppable Brilliance: Irish Geniuses and Asperger's Syndrome.* Dublin: Liberties Press.

Wallace, A. (1986) *The Prodigy.* New York: E. P. Dutton.

Weil, A. (1992) *The Apprenticeship of a Mathematician* (trans. J. Gage). Basle: Birkhäuser Verlag.

Welling, H. (1994) Prime number identificators in idiot savants: Can they calculate them? *Journal of Autism and Development Disorders* 24: 199–207.

Wertheimer, M. (1959) *Productive Thinking.* New York: Harper and Row.

Weyl, H. (1968) *Gesammelte Abbildungen* (ed. K. Chandrasekharan). Berlin: Springer Verlag.

Whitehead, A. N. (1911) *An Introduction to Mathematics.* London: Thornton Butterworth.

Wiener, N. (1953) *Ex-Prodigy: My Childhood and Youth.* New York: Simon and Schuster.

——. (1956) *I Am a Mathematician: The Later Life of a Prodigy.* New York: Doubleday.

Wigner, E. (1960) The unreasonable effectiveness of mathematics in the natural sciences. *Communications in Pure and Applied Mathematics* 13: 1–14.

Wilson, G. and Jackson, C. (1994) The personality of physicists. *Personality and Individual Differences* 16: 187–89.

Wimp, J. (1998) Review of books by A. C. Aitken. *Mathematical Intelligencer* 20: 62–79.

Wing, Lorna (1996) *The Autistic Spectrum.* London: Constable.

Wolff, S. (1995) *Loners.* London: Routledge.

Wolff, S. and Chess, S. (1964) A behavioural study of schizophrenic children. *Acta Psychiatrica Scandinavica* 40: 438–66.

Wolpert, L. (1988) *The Passion of Science.* Oxford: Oxford University Press.

Woolley, B. (1999) *The Bride of Science.* London: Macmillan.

Yandell, B. H. (2002) *The Honors Class: Hilbert's Problems and Their Solvers.* Natuck, MA: A. K. Peters.

Yates, F. and Mather, K. (1963) Ronald Aylmer Fisher. *Biographical Memoirs of Fellows of the Royal Society* 9: 91–120.

Young, D. A. B. (1994) Ramanujan's illness. *Notes and Records of the Royal Society of London* 48.

Young, F. A. (1967) Myopia and personality. *American Journal of Optometry* 44: 192–201.

Young, R. L. and Nettlebeck, T. (1994) The intelligence of numerical calculators. *American Journal of Mental Retardation* 99: 185–200.

Zuccala, A. (2005) *Revisiting the invisible college: A case study of the intellectual structure of singularity theory research.* Ph.D. thesis, University of Toronto.

Index